ELEMENTS OF DYNAMIC.

ELEMENTS OF DYNAMIC

AN INTRODUCTION TO THE STUDY OF

MOTION AND REST

IN SOLID AND FLUID BODIES

BY

W. K. CLIFFORD, F.R.S.

LATE FELLOW AND ASSISTANT TUTOR OF TRINITY COLLEGE, CAMBRIDGE;
PROFESSOR OF APPLIED MATHEMATICS AND MECHANICS AT
UNIVERSITY COLLEGE, LONDON.

PART I. KINEMATIC.

London:

MACMILLAN AND CO.

1878

Cambridge:

PRINTED BY C. J. CLAY, M.A.
AT THE UNIVERSITY PRESS.

CONTENTS.

BOOK I. TRANSLATIONS.

CHAPTER I. STEPS.

CHAPTER II. VELOCITIES.

CHAPTER III. CENTRAL ORBITS.

BOOK II.   ROTATIONS.

CHAPTER I.

CHAPTER II. VELOCITY-SYSTEMS.

CHAPTER III. SPECIAL PROBLEMS.

BOOK III. STRAINS.

CHAPTER I. STRAIN-STEPS.

CHAPTER II. STRAIN-VELOCITIES.

BOOK I. TRANSLATIONS.

CHAPTER I. STEPS.

INTRODUCTION.

JUST as Geometry teaches us about the *sizes* and *shapes* and *distances* of bodies, and about the relations which hold good between them, so Dynamic teaches us about the changes which take place in those distances, sizes, and shapes (which changes are called *motions*), the relations which hold good between different motions, and the circumstances under which motions take place.

Motions are generally very complicated. To fix the ideas, consider the case of a man sitting in one corner of a railway carriage, who gets up and moves to the opposite corner. He has gone from one place to another; he has turned round; he has continually changed in shape, and many of his muscles have changed in size during the process.

To avoid this complication we deal with the simplest motions first, and gradually go on to consider the more complex ones. In the first place we postpone the consideration of changes in size and shape by treating only of those motions in which there are no such changes. A body which does not change its size or shape during the time considered is called a *rigid* body.

The motion of rigid bodies is of two kinds; change of place, or *translation*, and change of direction or aspect, which is called *rotation*. In a motion of pure *translation*, every straight line in the body remains parallel to its original position; for if it did not, it would turn round, and there would be a motion of *rotation* mixed up with the motion of translation. By a *straight line in the body* we do not mean merely a straight line indicated by the shape or marked upon the surface of the body; thus if a box have a movement of translation, not only will its *edges* remain parallel to their original positions, but the same will be true of every straight line which we can conceive to be drawn joining any two points of the box.

When a body has a motion of translation it is found that every point of it moves in the same way; so that to describe the motion of the whole body it is sufficient to describe that of one point. When a body is so small that there is no need to take account of the differences in position and motion of its different parts, the body is called a *particle*. Thus the only motion of a particle that we take account of is the motion of translation of any point in it.

A motion of translation mixed up with a motion of rotation is like that of a corkscrew entering into a cork, and is called a *twist*.

Bodies which change their size or shape are called *elastic* bodies. Changes in size or shape are called *strains*.

The science which teaches how to describe motion accurately, and how to compound different motions together, is called *Kinematic* ($\kappa\iota\nu\eta\mu\alpha$, motion). We may conveniently reckon three branches of it, namely,

Kinematic of $\begin{cases} \text{Points or particles (Translations).} \\ \text{Rigid Bodies} \quad \text{(Rotations and Twists).} \\ \text{Elastic Bodies} \quad \text{(Strains).} \end{cases}$

It is found that the change of motion of any body depends partly on the position of distant bodies and partly on the strain of contiguous bodies. Considered as so depending, the rate of change of motion is called *force;* and the law just stated, expressing the circumstances under which motions change, is called the *law of force.*

The science which teaches how to calculate motions in accordance with the law of force is called *Dynamic* ($\delta\upsilon\nu\alpha\mu\iota\varsigma$, force). It is divided into two parts: *Static,* which treats of those circumstances under which *rest* or *null motion* is possible, and *Kinetic,* which treats of circumstances under which actual motion always takes place. Properly speaking, Static is a particular case of Kinetic which it has been convenient to consider separately.

When change of motion depends upon the position of distant bodies, it is also called *attraction* or *repulsion;* when it depends upon the strain of contiguous bodies, it is also called *stress.*

Those elastic bodies whose shape may change without stress (i.e., without simultaneous change of motion in adjacent bodies) are called *fluids;* all others are called *solids.* There are no known bodies whose size can change without stress.

The part of Dynamic which relates to fluid bodies is sometimes treated separately, under the name of *Hydro-dynamic* (Hydrostatic and Hydrokinetic).

That part which relates to the changes of shape of solid bodies, considered in relation to the law of force, is called the theory of *Elasticity.*

ON STEPS.

When a body has a motion of translation, all the points of it move along equal and similar paths. For let a and b be two points of the body, and let a move along the path $aa'a''$, and b along the path $bb'b''$, so that when a is at a', b is at b', and when a is at a'', b is at b''. Then, by the definition of a translation, the straight lines ab, $a'b'$ 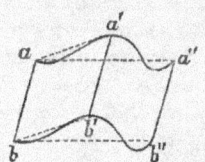 and $a''b''$ are equal and parallel. Consequently aa' is equal and parallel to bb', and aa'' to bb''. If therefore the path $aa'a''$ be moved so that a comes to b, and the lines aa', aa'', are kept parallel to their original positions, the points a', a'' must come to b', b'' respectively. But the point a' is *any* point on the path of a. Therefore every point on the path of a comes to coincide with the corresponding point on the path of b, or, which is the same thing, the path of a is equal and similar to the path of b. That is, the paths of *any two* points are equal and similar.

Hence it is sufficient, in describing the translation of a rigid body, to describe the motion of any one point of the body. But the former is really simpler than the latter; for the point starts from a definite place, which must be specified if its motion is fully described; but the fixing of this starting-point is unnecessary, as we have seen, when the motion of a point is only used to describe that of a rigid body.

At present we shall attend only to the *change of position* which a body undergoes between the beginning and end of the time considered, without troubling ourselves about what has taken place in the interval. That is, we shall pay attention to the fact that a has got to a' and b to b', without enquiring about the *paths aa'* and *bb'*, or about the *time* occupied in the transfer. A change of position effected by a motion of translation will be called a *step*.

The step of the point a from a to a' will be conveniently denoted by the symbol aa'; and we may represent it graphically by the straight line aa', provided we remember that the transfer takes place along any path whatever, and not necessarily along that straight line. This being so, the lines aa' and bb' will represent the *same* step of a rigid body if they are *equal in length* and *in the same direction;* that is, not merely parallel, but drawn in the same *sense* on two parallel straight lines. Thus a step of a rigid body is adequately represented by a line of given length and given direction drawn *anywhere*.

We shall say that the step aa' is equivalent to the step bb'; which may also be written shortly thus: $aa' = bb'$. Here the symbol $=$, which is commonly shorthand for *equal*, is used in the sense of *equivalent*. It means more than that the length aa' is equal to the length bb', namely, that the direction aa' is also the same as the direction bb'.

COMPOSITION OF STEPS. GEOMETRY.

If, while a railway carriage moves along the line from the position 1 to the position 2, a man who was sitting on the seat a moves across to the seat b, the final position of the man will be the same as the final position of b, namely, b'. The man is said to have made the step ab

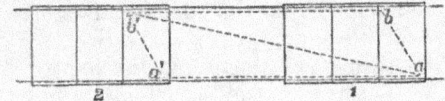

relative to the carriage; and his actual step from a to b' is said to be *compounded* of the step of the carriage, bb', and of this step relative to the carriage. Thus the step ab' is compounded of the steps ab and bb'. In this case ab and bb' are called the *components* and ab' is called the *resultant.*

Since aa' is equivalent to bb', we may equally speak of ab' as the resultant of aa' and ab. Thus we get two different rules for finding the resultant of two given steps :—

1. Let the straight lines representing the steps be so placed that the end of the first is the beginning of the second; then the step from the beginning of the first to the end of the second is the resultant (ab' resultant of ab and bb').

2. Let the straight lines representing the steps be so placed that they have the same beginning, and let a parallelogram be constructed of which they are two sides; then 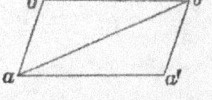 the resultant will be represented by that diagonal of the parallelogram which passes through the common beginning (ab' resultant of aa' and ab).

In the first rule we speak of the components as occurring in a certain order, first and second, viz., the step relative to the carriage and the step of the carriage; but in the second rule there is no such distinction. It appears from this that the two steps might be interchanged without affecting the result; and it is indeed obvious that if the train had moved sideways by the step ab, and the man had moved along it by the step aa', he would in the end be at b' as in the case already considered.

It is sometimes necessary to compound together more than two steps. Thus, in the example just used, the train is moving relatively to the Earth, the Earth is moving round the Sun, and the Sun is moving on his own account through space,—or rather, for this is all we can be sure about, he is moving relatively to certain stars. So that to get the actual motion of the man in the train relative to these stars, we must compound all these motions together. The rule for this is very easily found when the straight lines representing the steps to be compounded are so arranged that the end of each is the beginning of the next. Then the resultant is the step from the beginning of all to the end of all.

Thus the steps ab, bc, cd, de have the resultant ae; for ab and bc give ac, then ac and cd give ad, and finally ad and de give ae.

But when the lines are all arranged so as to have a common beginning, the rule is rather more complex, and will be examined after we have found a shorter way of writing about the composition of steps.

What is true of two steps, that their resultant is independent of the order in which they are taken, is true of any number of steps. This we shall now prove.

First, *the resultant is unaltered by the interchange of two successive steps.* For to interchange the steps bc, cd, that is, to take cd before bc, we must draw bc' equal and parallel to cd, and then from c' a line equal and parallel to bc. But this line will end precisely at d, because $bcdc'$ is a parallelogram. Nothing after the point d will be altered, and consequently the resultant ae will be the same as before.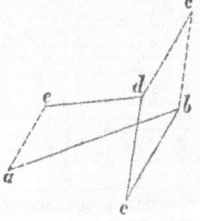

Next, *any change whatever in the order can be produced by a sufficient number of interchanges of successive steps.* This statement clearly does not apply to steps only, but to any things whatever that can be arranged in order; for example, letters or figures. The truth of the statement will be made clear by an example of the process to be used. Thus, let it be required to change the order 123456 into the order 314625. Bring 3 to the first place by successively interchanging it with 2 and 1. Then 1 will be in the second place as required. Bring 4 to the third place by interchanging it with 2, and then bring 6 to the fourth place by interchanging it with 5 and 2; lastly, interchange 5 and 2, and the required transformation is complete.

As no one of these six interchanges has altered the resultant, it remains the same as at first. Thus the proposition is proved.

COMPOSITION OF STEPS. ALGEBRA.

When we have to deal with steps which are all in the same straight line, as ab, bc, cd, we may describe each of them as a step of so many inches to the right or to the left. To find the resultant we must add together the lengths of all the steps to the right, and also the lengths of all the steps to the left. The resultant is a step whose length is the difference between these two sums, and it is to the right if the former is greater, to the left if the latter is greater. Thus the resultant of the steps ab, bc, cd is, as we know, ad; and the length of ad is $ab + cd - cb$. The resultant is a step to the right because the sum $ab + cd$ is greater than cb.

It is convenient to regard a step to the left as a negative quantity, the *addition* of which is equivalent to the *subtraction* of its length from that of a step to the right. Thus $+ bc$ is taken to be the same as $- cb$. And thus we may write either

$$ad = ab + cd - cb,$$

or else

$$ad = ab + cd + bc.$$

The symbol $+$, placed between two steps, is thus made to mean that their resultant is to be found, regard being had to their directions. The resultant $ab + bc$ is always ac, no matter how the points are situated; but the *length* ac is a sum or a difference of the lengths ab and bc, according as they are in the same direction or not.

We shall extend this meaning of the symbol $+$ to cases in which the component steps are not in the same.

straight line; that is to say, $ab + cd$ shall always mean
the resultant of the steps ab and cd, not the sum of
their lengths unless this is expressly mentioned. Similarly $ab - cd$ will mean the resultant of ab and a step the
reverse of cd, namely dc.

After a little practice, the student will find that this
extension of the meaning of the signs $+$, $-$, and $=$ does
not cause any confusion, but on the contrary enables us to
reason more clearly because more compactly. We shall
now use this method to investigate the resultant of several steps the lines representing which are so placed as all
to have the same beginning.

In the case of two steps oa and ob,
the rule is to complete the parallelogram $oapb$, and then the diagonal op
is the resultant. But if we join the
points ab by a straight line meeting
op in c, both op and ab are bisected at the point c. Thus
op is twice oc, which may be written $op = 2oc$. Observe
that $2oc$ means a step in the direction of oc, of twice its
length. We may now state our rule as follows:—find c
the middle point of ab, then the resultant of the steps
oa and ob is twice oc; or, more shortly, $oa + ob = 2oc$.

We may extend this result. Let
ab be divided in c so that ac is to cb
as m to l, where l and m are any two
numbers. Then

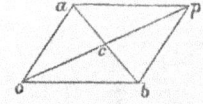

$$l . ac = m . cb \text{ and } (l + m) . ac = m . ab.$$

Now $oc = oa + ac$; that is to say, the
step oc is the resultant of oa and ac. Therefore

$$(l + m) oc = (l + m) oa + (l + m) ac.$$

But $(l + m) ac = m . ab$, and $ab = ao + ob = ob - oa$.
Substituting this value, we find

$$(l + m) oc = (l + m) oa + m . ab$$
$$= (l + m) oa + m (ob - oa)$$
$$= l . oa + m . ob.$$

That is, *if ab be divided in the ratio $m : l$ at the point c,
then the resultant of l times oa and m times ob is $l + m$
times oc.*

We shall now write this proof in a shorter form. We have
$$oa = oc + ca,$$
$$ob = oc + cb \; ;$$
therefore
$$l \cdot oa + m \cdot ob = (l + m) \cdot oc + l \cdot ca + m \cdot cb$$
$$= (l + m) \cdot oc,$$
because the point c was so chosen that $l \cdot ac = m \cdot cb$, or (which is the same thing),
$$l \cdot ca + m \cdot cb = 0.$$

The former investigation exhibits the process of finding oc in terms of oa and ob; the latter is a shorter and more symmetrical proof of the result when it is known.

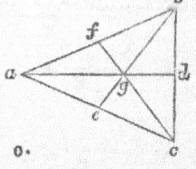

We proceed now to the case of three steps, oa, ob, oc. Bisect ab in f, then $2of = oa + ob$, so that it remains to find the resultant of oc and twice of. This is a case of the last proposition, in which $l = 1$ and $m = 2$; we must therefore divide cf in the ratio of $2 : 1$. Taking then a point g at two-thirds of the way from c to f, we find $3og$ for the resultant of oa, ob, oc; or, more shortly,
$$oa + ob + oc = 3og.$$

This result is true wherever the point o is: whether in the plane abc or out of it. And the method of determining g is quite independent of the position of o. By making o coincide with g, so that og is zero, we find that $ga + gb + gc = 0$. This is independently clear, because $ga + gb = 2gf$, and $2gf + gc = 0$ by construction. Hence also we see that g is $\frac{2}{3}$ of the way from a to d, and from b to e, if d, e are the middle points of bc, ca. Or the lines joining the angles of a triangle to the middle points of its sides meet in a point which divides each of them in the ratio of 2 to 1.

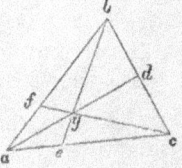

To find the resultant of l times oa, m times ob, and n times oc, we must observe that whatever the point g is,
$$oa = og + ga, \quad ob = og + gb, \quad oc = og + gc,$$
and therefore
$$l \cdot oa + m \cdot ob + n \cdot oc = (l + m + n) \cdot og$$
$$+ l \cdot ga + m \cdot gb + n \cdot gc.$$

If therefore we can find a point g such that
$$l \cdot ga + m \cdot gb + n \cdot gc = 0,$$
we shall have $l \cdot oa + m \cdot ob + n \cdot oc = (l + m + n) \cdot og$. Now $l \cdot ga + m \cdot gb = (l + m) \cdot gf$, if f is the point dividing ab in the ratio $m : l$. Hence $(l + m) \cdot gf + n \cdot gc = 0$, or g is the point dividing cf in the ratio $l + m : n$. We might equally well have found g by dividing bc in the ratio $n : m$ at d, or ca in the ratio $l : n$ at e. That is, we have the equations $l.fa = m.bf$, $n.ec = l.ae$, and $m.db = n.cd$, between the lengths of the six segments into which the sides are divided. Multiplying these equations together, we find that the product lmn divides out, and that $fa \cdot ec \cdot db = bf \cdot ae \cdot cd$. Hence *if ad, be, cf meet in a point, then af.ce.bd = ea.dc.fb.* This theorem is a useful criterion for the concurrence of three lines drawn through the vertices of a triangle.

A similar set of theorems belongs to the composition of *four* steps. If f, g, h are the middle points of bc, ca, ab, and f', g', h' of da, db, dc, then ff', gg', hh' bisect one another at a point k, such that

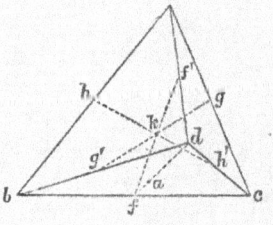

$$oa + ob + oc + od = 4ok.$$
For $oa + ob = 2oh$,
and $oc + od = 2oh'$;

also if k be taken at the middle point of hh', $oh + oh' = 2ok$; therefore $oa + ob + oc + od = 2 (oh + oh') = 4ok$. And the symmetry shews that this k is also the middle point of gg' and of ff'.

Moreover, if we take α ⅓ of the way from f to d, then k is ¼ of the way from α to a. For we know that $ob + oc + od = 3o\alpha$, and therefore $oa + 3o\alpha = 4ok$, wherever o is: or taking o to coincide with k, $ka + 3k\alpha = 0$, which shews that k divides $a\alpha$ in the ratio of $3 : 1$.

Observe that the points $abcd$ may either be in the same plane, or form a triangular pyramid, or *tetrahedron.*

DESCRIPTION OF STEPS. 11

In general, if we have n steps oa_1, oa_2, $oa_3...oa_n$, it is always possible to find a point g such that

$$n \cdot og = oa_1 + oa_2 + ... + oa_n = \Sigma oa,$$

as this sum may be conveniently written. The position of the point g will depend upon the points a_1, $a_2...a_n$, but not in the least upon the point o. To prove this, suppose we take a point p, and draw the steps pa_1, $pa_2...pa_n$. The resultant of these must be some step, which can be found by arranging them tandem as in our first process. Let pg be the nth part of this resultant, so that $n \cdot pg = \Sigma pa$. Now we know that

$$og = op + pg, \quad oa_1 = op + pa_1, \quad ... oa_n = op + pa_n.$$

Therefore $n \cdot og = n \cdot op + n \cdot pg = n \cdot op + \Sigma pa = \Sigma oa$.

Thus g being chosen so that $n \cdot pg = \Sigma pa$ for a *particular* position of p, we see that $n \cdot og = \Sigma oa$ for any point o whatever. This point g is called the *mean point*, or *mid-centre*, of the points a_1, $a_2...a_n$.

Similarly, it may be shewn that there is a point g such that, if l_1, $l_2...l_n$ are any numbers,

$$(l_1 + l_2 + ... + l_n) \, og = l_1 \cdot oa_1 + l_2 \cdot oa_2 + ... + l_n \cdot oa_n,$$

whatever point o is.

And lastly, if we have n steps a_1b_1, $a_2b_2...a_nb_n$, anyhow situated, their resultant is n times the step from the mean point of a_1, $a_2...a_n$ to the mean point of b_1, $b_2...b_n$. The proof of this is left as an exercise for the reader.

RESOLUTION AND DESCRIPTION OF STEPS.

We have already seen that a step in a known direction may be completely specified by describing its length. This may be done in two ways. First, *approximately*, by stating the number of inches or centimeters and parts of an inch or centimeter; if the parts are expressed in decimal fractions, the approximation may be carried to any required degree of accuracy by taking a sufficient number of places of decimals. But as the length to be described is generally incommensurable in regard to an inch or a centimeter, this method is very rarely anything *more* than an approximation. Second, *graphically*, by *drawing the length to scale.* A certain line being marked out upon the diagram to represent a centimeter, another line is drawn bearing the same ratio to this one that the length to be described bears to a centimeter. Thus at the side of a map there is a *scale of miles*, by which the distance between two places may be estimated. The actual distance bears the same ratio to a mile that the distance on the map bears to the representative length on the scale. This is the theoretically correct way of representing all continuous quantities, except angles, which should also be drawn; though it is sometimes convenient to describe an angle in terms of degrees, minutes and seconds; or in *circular measure*, which is the ratio of its arc to the radius.

When it is known that a step lies in a certain plane, it may always be resolved into two components which are in fixed directions at right angles to one another. Let oX, oY be two fixed lines at right angles to one another. Let op be the step which it is required to resolve. Draw

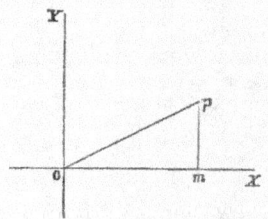

pm perpendicular to oX, then $op = om + mp$; or the step op has been resolved into two, one of which is in the direction oX, and the other in the direction oY.

Let x be the number of units of length (e.g. centimeters) in om, and y the number in mp. Let also i represent a step of one centimeter along oX, and j a step of one centimeter along oY. Then om is x times i, or xi; and mp is y times j, or yj. Hence the step $op = xi + yj$; and we may say that *every step in the given plane may be described in the form $xi + yj$, where x and y are two numerical ratios, and i, j are fixed unit steps at right angles to one another.*

When the lengths x and y are given either approximately or graphically, the step (known to lie in a given plane) is completely described in the same way.

It is to be understood that when m falls to the *left* of

oY, x is a negative quantity; and when p falls *below* oX, y is a negative quantity.

When it is not known in what plane a step lies, we can still resolve it into *three* components along fixed directions at right angles to one another. Let oX, oY, oZ be three lines at right angles to one another, op the step to be resolved. Draw pn perpendicular to the plane XoY, and nm perpendicular to oX. Then $op = om + mn + np$, or the step op has been resolved into three, which are respectively in the directions oX, oY, oZ.

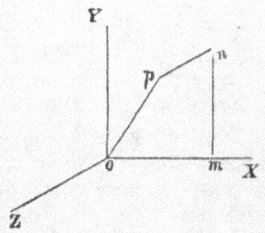

Let, as before, x, y be the number of centimeters in om, mn, and let z be the number in np. Let also i, j, k be three steps of one centimeter each in the directions oX, oY, oZ. Then $om = xi$, $mn = yj$, $np = zk$, and $op = xi + yj + zk$. Thus we see that *any step whatever can be described in the form* $xi + yj + zk$, *where* x, y, z *are three numerical ratios, and* i, j, k *are fixed unit steps at right angles to one another.*

When the lengths x, y, z are given approximately or graphically, the step is completely described in the same way. It is to be understood that z is reckoned negative when p lies on the *further* side of the plane XoY.

We shall find other quantities, besides steps, which can be resolved into components in three fixed directions, and completely described by assigning three lengths. All such quantities are called *vectors*, or *carriers*, from their analogy to a step of translation or carrying. They can always be described in the form $xi + yj + zk$, where i, j, k are fixed unit vectors at right angles to one another. Except these unit vectors, it is usual to represent a vector either by the beginning and end of the line representing it, as op, or by a single small Greek letter, as α, ρ.

When the position of a point p is described by means of the step from a fixed point o to it, the point o is called the *origin*, and the components x, y, z are called the *co-ordinates* of p. The lines oX, oY, oZ are called *axes of co-ordinates,* and the planes which contain them in pairs the *co-ordinate planes.* The step or vector *op* is called the *position-vector* of the point *p.*

REPRESENTATION OF MOTION.

We go on to describe more completely the translation of a rigid body. Hitherto we have considered only the step from the beginning to the end of the motion; we shall now take account of the path and of the time in which it is described. As before it will be sufficient to consider the motion of a single point of the body.

To describe completely the motion of a point *p* from *a* to *b* it would be necessary to assign the path and also the position of the point in the path at every instant of time. The path may be assigned by drawing it, or by stating its geometrical properties. The position of the point in the path may be assigned by giving the length *ap* measured along the path at every instant; and this may be done in two ways.

First, by the approximate or numerical method. We may construct a table, in the first column of which are marked seconds or fractions of a second, and in the second are written against them the number of centimeters in the length *ap* at that time. Tables on this principle are printed in the *Nautical Almanac,* giving the position at

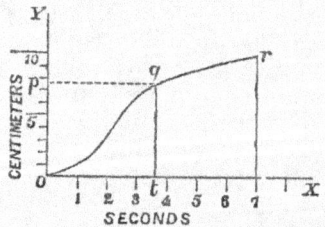

any time of the Sun and the planets; principally of the Moon. The method is imperfect, because it only gives the position at certain selected moments, and then only approximately.

Secondly, by the graphical method. In this, the seconds are marked off on a horizontal line oX, and above every point of this there is set up a straight line representing the distance traversed at that instant. Thus, at the instant t, about $3\frac{1}{2}$ seconds from the beginning of the motion, the distance traversed was tq, on the scale of centimeters marked on oY. Drawing qp horizontal to meet oY, we find the distance about $7\frac{1}{2}$ centimeters.

The tops of all these lines form a curve oqr, which is called the *curve of positions* of the moving point. The figure is equivalent to a table with an infinite number of entries, each of which is exact. The line oX is the first column, and the lengths tq, etc., answer to the second column.

In certain ideal cases of motion, it is possible to get rid of one objection to the numerical method, and to make it partially describe the position of the point at every instant of time. This is when we can state a *rule for calculating* the number of centimeters passed over from the number of seconds elapsed; or, which is the same thing, when we can find an algebraical formula which expresses the distance traversed in terms of the time. Such motions do not occur accurately in nature; but there are natural motions which closely approximate to them, and which for practical purposes are adequately described in this way. We go on to consider some of these ideal motions.

UNIFORM MOTION.

When equal distances are gone over in equal times, the motion is said to be *uniform*.

In uniform motion, the distances gone over in unequal times are proportional to the times (Archimedes). For let t and T be unequal times in which the distances s and S are gone over. Take any two whole numbers m and n. Then if we take n intervals of time equal to t, there will be gone over in them n distances equal to s; that is, a distance ns is gone over in the time nt. Similarly, mS will be gone over in the time mT. Now if nt is greater than mT, ns is greater than mS; for in a greater time a greater distance must be traversed. If nt is equal to mT, ns is equal to mS; and if nt is less than mT, ns is less than mS. Hence by Euclid's definition of proportion,

$$S \,:\, s = T \,:\, t.$$

Let v be the number of centimeters gone over (or *described*) in one second; then $s \,:\, v = t \,:\, 1$, or $s = vt$, where s is the number of centimeters described in t seconds. Here all three numbers may be incommensurable; but the algebraic formula $s = vt$ supplies us with a rule for calculating s when t is known; viz., multiply t by v.

The curve of positions in this case is a straight line. For, if we set up the length v above the point 1, and draw through o the straight line ovq; then on drawing tq vertical through any point t, we shall have $tq \,:\, v = ot \,:\, 1$, or tq correctly represents the distance described in the time ot.

Uniform motion may of course take place along any path whatever. But there are two cases of special interest; when the path is a *straight line* and when it is a circle.

UNIFORM RECTILINEAR MOTION.

Let p be a point moving uniformly along the straight line abp, and let o be any fixed point. We shall completely describe the position of the point p at any instant, if we specify the step which must be taken to go to p from o at that instant. Now $op = oa + ap$. Let ab be the distance traversed in one second, then ap, being the distance traversed in t seconds, is $t \,.\, ab$. Hence we have

$$op = oa + t \,.\, ab,$$

or, if we denote the step op by ρ, oa by α, ab by β, then

$$\rho = \alpha + t\beta.$$

This is called the *equation* of uniform rectilinear motion. It is simply shorthand for this statement:—the steps to be taken in order to get from o to the position of p after t seconds are, first, the step α (oa) which takes us to the position at the beginning of the motion, and then t times the step β (ab).

Two uniform rectilinear motions compound into a uniform rectilinear motion.

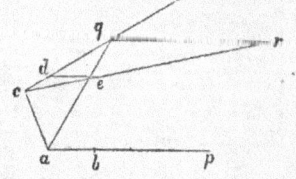

While p moves uniformly along the line ab, let q move uniformly, relative to p, along cd; and let cd be the distance traversed in one second in the relative motion. Draw de equal and parallel to ab, then ce is the actual motion of q in one second. Draw qr parallel to ab, meeting ce produced in r. Then, cq being traversed in the same time as ap, we must have $cq : cd = ap : ab = t : 1$. Now $cq : cd = qr : de$, so that $qr = ap$. Hence r is the actual position of q at the end of the time t. It is in the straight line ce, and $cr : ce = cq : cd = t : 1$. Thus the actual motion of q is a uniform rectilinear motion.

The same thing appears by considering the equations. Let ρ_1 be the step op, and ρ_2 the step pq; then $\rho = \rho_1 + \rho_2$ is the step oq. Now we have

$$\rho_1 = \alpha_1 + t\beta_1, \quad \text{where} \quad \alpha_1 = oa, \quad \beta_1 = ab,$$
$$\rho_2 = \alpha_2 + t\beta_2, \qquad \text{,,} \qquad \alpha_2 = ac, \quad \beta_2 = cd,$$

and therefore

$$\rho = \rho_1 + \rho_2 = \alpha_1 + \alpha_2 + t(\beta_1 + \beta_2),$$

the equation to a uniform rectilinear motion.

The curve of positions of any motion whatever may be conceived to be constructed by help of a uniform rectilinear motion, in this way. Let the original motion be that of a point p along the path ab:

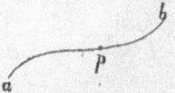

Let a point p' move along oY at the
same time, so that the distance op'
is at every instant equal to the distance ap measured along the path.
While this motion takes place, let the
straight line oY have a uniform horizontal translation of one centimeter
in every second; then by this com-

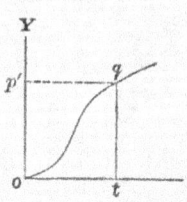

bination of motions the point p' will describe the curve of
positions oq.

Hence *the curve of positions of any rectilinear motion
is described by combining that motion with a uniform
rectilinear motion of one centimeter per second in a direction at right angles to it.*

UNIFORM CIRCULAR MOTION.

In uniform circular motion every point p of the moving
body goes round a circle so as to describe equal arcs in
equal times, and therefore proportional arcs in different
times.

The radius of the circle is called the *amplitude* of the
motion.

The time of going once round is called the *period*.

If the arcs measured on the circle are reckoned from a
point a, and if the moving point started from e at the
beginning of the time considered, the angle aoe is called
the *angle at epoch*, or shortly the *epoch*. Strictly speaking, the *epoch* is the beginning of the time considered.

The ratio of the arc ap to the whole
circumference is called the *phase* at any
instant.

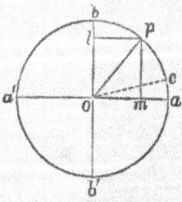

Let n be the circular measure of the
arc described in one second, and a the
radius of the circle; so that na is the
length of the arc described in one
second. Then nat is the length of arc,
ep, described in t seconds, and nt is its circular measure.

Let also ϵ be the circular measure of aoe; then circular measure of aop

$$= nt + \epsilon.$$

We shall now obtain an expression for the step op at any instant. Draw pm, ob, perpendicular to oa. Then $op = om + mp$. Now as far as lengths are concerned, $\dfrac{om}{op} = \cos aop$, and $\dfrac{mp}{op} = \sin aop$. Or, since $op = oa = ob$ in length, $om = oa \cos aop$ and $mp = ob \sin aop$. In the equation $om = oa \cos aop$, the quantities om and oa may be regarded as steps; for as they are in the same direction, one is equal to the other multiplied by the numerical ratio $\cos aop$. The same may be said of the equation $mp = ob \sin aop$. Now $aop = nt + \epsilon$, and therefore

$$op = oa \cdot \cos(nt + \epsilon) + ob \cdot \sin(nt + \epsilon),$$

or if we write ρ for op, ai for oa, and aj for ob, so that i, j are unit steps along oa, ob, then

$$\rho = a\{i \cos(nt + \epsilon) + j \sin(nt + \epsilon)\}.$$

This is the equation to uniform circular motion. The angle $nt + \epsilon$ is called the *argument* of this expression for ρ.

A circular motion which goes round like the hands of a clock, or *clockwise*, is said to be *in the negative sense*; one that goes round the other way, or *counter-clockwise*, is said to be *in the positive sense*.

Two uniform circular motions of the same period and the same sense compound into a uniform circular motion of that period and sense.

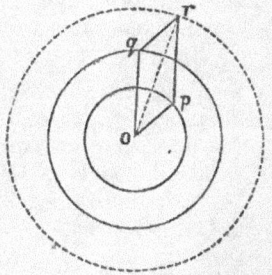

Suppose the circles so placed as to have the same centre. The motions of p and q relative to o may be combined by completing the parallelogram $oprq$; then the motion of r is the resultant. We may consider the parallelogram $oprq$ to be made of four jointed rods, of which op and oq turn round o. When these motions have the

2—2

same period and the same sense, the angle poq remains always constant; therefore the shape of the parallelogram remains unchanged. Consequently or is of constant length, and makes always the same angle with op or oq. Hence r goes round uniformly in a circle of radius or.

Let $op = \rho_1$, $oq = \rho_2$, $or = \rho$. Then, if a, b are the amplitudes, i, j unit steps at right angles to one another,

$$\rho_1 = ai \cos (nt + \epsilon_1) + aj \sin (nt + \epsilon_1),$$

$$\rho_2 = bi \cos (nt + \epsilon_2) + bj \sin (nt + \epsilon_2),$$

$$\rho = \rho_1 + \rho_2 = i\{(a\cos\epsilon_1 + b\cos\epsilon_2)\cos nt - (a\sin\epsilon_1 + b\sin\epsilon_2)\sin nt\}$$

$$+ j\{(a\cos\epsilon_1 + b\cos\epsilon_2)\sin nt + (a\sin\epsilon_1 + b\sin\epsilon_2)\cos nt\},$$

which may be written

$$ci \cos (nt + \epsilon) + cj \sin (nt + \epsilon),$$

provided that $\quad a \cos \epsilon_1 + b \cos \epsilon_2 = c \cos \epsilon,$

$$a \sin \epsilon_1 + b \sin \epsilon_2 = c \sin \epsilon.$$

From these two equations we must find c and ϵ. Dividing the second by the first, we find

$$\tan \epsilon = \frac{a \sin \epsilon_1 + b \sin \epsilon_2}{a \cos \epsilon_1 + b \cos \epsilon_2}.$$

Squaring both sides of both equations, and adding them together, we find

$$c^2 = a^2 + b^2 + 2ab \cos (\epsilon_1 - \epsilon_2).$$

These formulæ determine the amplitude and epoch of the resultant motion. It is left to the reader to verify them by comparison with the geometrical solution.

Like the preceding theorem about uniform rectilinear motions, this theorem may be extended to any number of circular motions of the same period and sense; by first compounding the first two, then the third with their resultant, and so on. Or the extended theorem may be proved directly, either by the geometrical or by the analytical method.

HARMONIC MOTION.

While the point p moves uniformly round a circle, let a perpendicular pm be continually let fall upon a diameter

aa'. Then the point m will oscillate to and fro between a and a'. This motion of the point m is called *simple harmonic motion*.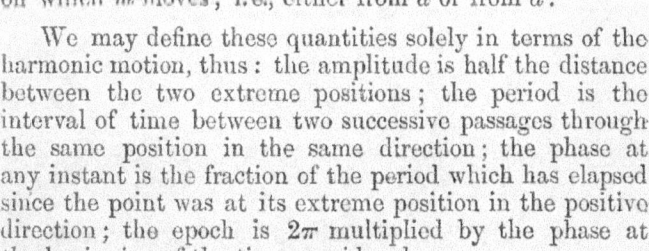

The *amplitude, period, epoch*, and *phase* of the simple harmonic motion are the same as those of the uniform circular motion of p. The epoch, however, must be reckoned from one extremity of the diameter on which m moves; i.e., either from a or from a'.

We may define these quantities solely in terms of the harmonic motion, thus : the amplitude is half the distance between the two extreme positions ; the period is the interval of time between two successive passages through the same position in the same direction ; the phase at any instant is the fraction of the period which has elapsed since the point was at its extreme position in the positive direction ; the epoch is 2π multiplied by the phase at the beginning of the time considered.

The *equation* to the simple harmonic motion is

$$om = oa \cos aop = oa \cos (nt + \epsilon) ;$$

or $$\rho = \alpha \cos (nt + \epsilon).$$

Here the amplitude is α, the period is $\dfrac{2\pi}{n}$ (for since in time t the circular measure of the arc described is nt, it follows that in time $\dfrac{2\pi}{n}$ the circular measure is 2π), the epoch is ϵ, and the phase is $\dfrac{nt + \epsilon}{2\pi}$.

Uniform circular motion is compounded of two simple harmonic motions of equal period, whose amplitudes are equal in length and perpendicular in direction, and whose phases differ by $\frac{1}{4}$. Namely, the motion of p is compounded of the motions of l and m, which answer to this description. *Any* two diameters at right angles will serve for this resolution.

The curve of positions of a simple harmonic motion may be constructed by means of a right circular cylinder. (This surface is traced out by a straight line which revolves about a fixed parallel line ; the moving line is called a *generator*, the fixed line the *axis*, of the cylinder.) Cut the cylinder through obliquely by a plane cc', and through o the centre of cc' draw a plane perpendicular to the axis of the cylinder, which will cut the cylinder in a circle $aba'b'$. Let bb' be the intersection of the two planes. A

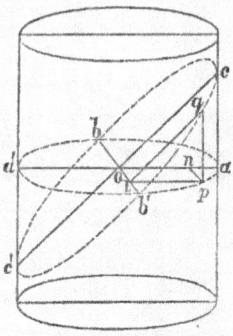

plane through o perpendicular to bb' will contain cc' and aa', and everything will be symmetrical in regard to this plane.

The curve in which the plane cc' cuts the cylinder is called an *ellipse.* We shall shew that if a piece of paper be wrapped round the cylinder, marked along this curve, and afterwards unrolled and laid flat, the trace upon it will be the curve of positions of a simple harmonic motion*. Let q be a point on this curve; draw qp per-

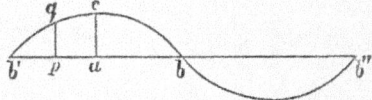

pendicular to the plane $aba'b'$, meeting the circle in p; draw pn, pl perpendicular to aa' and bb' respectively. Then the triangle qpl is similar to cao. Therefore

$$pq : lp = ac : oa, \quad \text{or} \quad pq = ac \cdot \frac{on}{oa} = ac \cos aop.$$

If then p moves uniformly round the circle $aba'b'$ at the rate of one centimeter per second, we shall have

$$pq = ac \cos (nt + \epsilon), \quad \text{where} \quad n \cdot a = 1.$$

* The reader should cut out in paper a wavy curve of the shape drawn in the figure, and then bend it into the form of a cylinder, when the plane elliptic section will become manifest.

Hence pq will at every instant be the step from its mean position to a point which is moving in a simple harmonic motion of amplitude ac, period $2\pi . oa$. When therefore the figure is unrolled from the cylinder, the wavy curve (called the *harmonic curve*, or *curve of sines* because the ordinate pq is equal to $ac . \sin \dfrac{\pi}{2} \dfrac{b'p}{b'a}$, that is, proportional to the *sine* of a multiple of the abscissa $b'p$) is the curve of positions of the simple harmonic motion aforesaid. The amplitude is the height of a wave, ac. The period is the *length* of a wave, $b'b''$, every centimeter in that length representing a second of time.

The curves of position of motions compounded of simple harmonic motions in one line may be constructed by actually compounding the curves of position of the several motions—that is, by adding together their ordinates to form the ordinate of the compound curve. Thus in the figures the height of the dark curve above the horizontal line is at every point *half* the algebraic sum (which is more convenient for drawing than the whole sum) of the heights of the other two; as for example $2mq = mp + mr$. A depth below the line is counted as a negative height. The first figure represents the composition of two simple harmonic motions of the same period; the second two such motions in which the period of one is half that of the other. The student should construct a series of these for different epochs of one of the motions, and then compare them with those figured in Thomson and Tait's *Natural Philosophy*, p. 43.

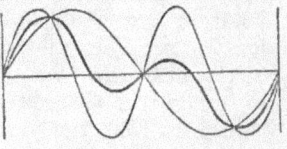

In the case where the component motions have the same period, the resultant is a simple harmonic motion of that period. This follows at once from the corresponding theorem in regard to circular motions. Completing the parallelogram $opqr$, and drawing perpendiculars pl, qm, rn

upon aa', we see that $ol = qs$ $= mn$, and consequently $on = ol$ $+ om$. Therefore the motion of n is compounded of the motions of l and m. But since r moves uniformly in a circle, the motion of n is a simple harmonic motion. And we have seen that, when

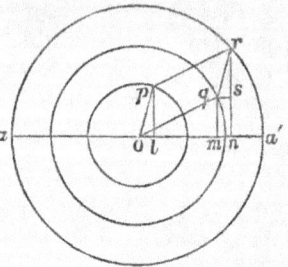

$$\rho_1 = a \cos (nt + \epsilon_1),$$

$$\rho_2 = b \cos (nt + \epsilon_2),$$

then $\qquad \rho = \rho_1 + \rho_2 = c \cos (nt + \epsilon),$

provided that $\quad c^2 = a^2 + b^2 + 2ab \cos (\epsilon_1 + \epsilon_2),$

and $\qquad \tan \epsilon = \dfrac{a \sin \epsilon_1 + b \sin \epsilon_2}{a \cos \epsilon_1 + b \cos \epsilon_2}.$

It follows at once that the theorem is true for any number of simple harmonic motions having the same period.

The use of the jointed parallelogram $opqr$ for compounding harmonic motions of different periods is exemplified in Sir W. Thomson's Tidal Clock. The clock has two hands whose lengths are proportional to the solar and lunar tides respectively, while their periods of revolution are equal to the periods of those tides. A jointed parallelogram is constructed, having the hands of the clock for two sides. If the clock is properly set, the height of that extremity of the parallelogram which is furthest from the centre will be continually proportional to the height of the compound tide. For this purpose a series of horizontal lines at equal distances is drawn across the face of the clock, and the height is read off by running the eye along these to a vertical scale of feet in the middle.

ON PROJECTION.

The foot of the perpendicular from a point on a straight line or plane is called the *orthogonal projection* of that point on the line or plane, or more shortly (when no mistake can occur) the *projection* of the point. Thus the point m is the projection of p on the straight line aa'. We say also, by a natural extension, that the *motion* of m is the projection of the motion of p. Thus simple harmonic motion is the orthogonal projection of uniform circular motion on any straight line in the plane of the circle.

When all the points of a figure are projected, the figure formed by their projections is called the projection of the original figure. Thus, for example (first figure of p. 22), the circle $aba'b'$ is the projection of the ellipse $cbc'b'$, for it is produced by drawing perpendiculars from every point of the ellipse to the plane. The point a is the projection of c, a' of c', p of q, etc.; b and b' are their own projections, being already in the plane of the circle.

Instead of drawing lines *perpendicular* to a plane from all the points of a figure, we may also project it by drawing lines all parallel to one another, but in some other direction. This is called *oblique projection*. The ellipse $cbc'b'$ is an oblique projection of the circle $aba'b'$, for the lines ac, $a'c'$, pq are all parallel to one another, although they are not at right angles to the plane of the ellipse. Orthogonal and oblique projections are both included under the name *parallel projection*, because in both cases the projection is made by drawing lines which are all parallel to one another.

We may also project a figure on to a given plane by means of lines drawn through a fixed point; this is called *central* projection. It occurs whenever a *shadow* is cast by a luminous point. If we suppose the centre of projection c to move away to an infinite distance, the lines converging to it will all become parallel. Thus we see that parallel projection is only a particular case of central projection in which the centre of projection has gone away to an infinite distance. The shadow cast by a bright star is for all practical purposes a parallel projection.

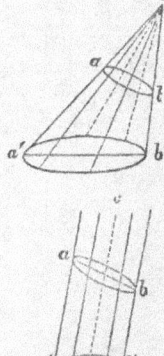

The projection of a straight line is made by drawing
a plane through it and through the centre of projection.
Thus if we draw the plane cab and produce it to meet
the plane of projection in $a'b'$, this line $a'b'$ will be the
projection of ab. In parallel projection we must draw
through the line a plane parallel to the projecting lines,
like the plane $ab\,a'b'$ in the second figure. We see in
this way that the projection of a straight line is always a
straight line, and that, since the line and its projection
are in the same plane, they must either meet at a finite
distance or be parallel (meet at an infinite distance).

*In parallel projection, parallel lines are projected into
parallel lines, and the ratio of their lengths is unaltered.*
Through the parallel lines ab, cd
we must draw the planes $ab\,a'b'$,
$cd\,c'd'$ both parallel to the projecting lines, and therefore
parallel to each other. These
planes will consequently be cut
by the plane of projection in
the parallel lines $a'b'$, $c'd'$.
Moreover the triangles pbb',
qdd', having their respective

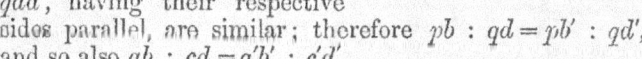

sides parallel, are similar; therefore $pb : qd = pb' : qd'$,
and so also $ab : cd = a'b' : c'd'$.

*The orthogonal projection of a finite straight line on a
straight line or plane is equal in length to the length of the
projected line multiplied by the cosine of its inclination to
the straight line or plane.* If pq is the
projection of PQ, draw pq' equal and
parallel to PQ. Then Qq' is parallel to
Pp and therefore perpendicular to pq;
therefore the plane Qqq' is perpendicular to pq, and therefore $q'q$ is perpendicular to pq. Hence $pq = pq'\cos$
$q'pq = PQ \times$ cosine of angle between
PQ and pq.

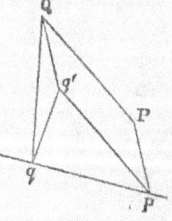

*The orthogonal projection of an area on a plane is equal
to the area multiplied by the cosine of its inclination to the*

plane. This is clearly true for a rectangle *ABCD*, one of whose sides is
parallel to the line of intersection of
the planes. For the side *AB* is unaltered, and the other, *BC*, is altered
into *Bc*, which is *BC* cos *θ*. Hence it
is true for any area which can be made
up of such rectangles. But any area
A can be divided into such rectangles
together with pieces over, by drawing
lines across it at equal distances perpendicular to the intersection of the
two planes, and then lines parallel to
the intersection through the points
when they meet the boundary. All

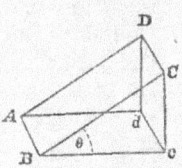

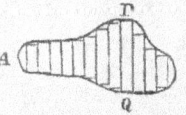

these pieces over, taken together, are less than twice
the strip whose height *PQ* is the difference in height
between the lowest and highest point of the area; for
those on either side of it can be slid sideways into that
strip so as not to fill it. And by increasing the number
of strips, and diminishing their breadth, we can make
this as small as we like. Let then *A'* be the sum of
the rectangles, then *A'* can be made to differ from *A*
as little as we like. Now the projection of *A'* is *A'* cos *θ*,
and this can be made to differ from the projection of
A as little as we like. Therefore there can be no finite
difference between the projection of *A* and *A* cos *θ*, because *A'* cos *θ* can be made to differ as little as we like
from both of them.

PROPERTIES OF THE ELLIPSE.

The ellipse may be defined in various ways, but for
our purposes it is most convenient to define it as the
parallel projection of a circle. This definition leads
most easily to those properties of the curve which are
chiefly useful in dynamic.

Centre. The centre of a circle bisects every chord
passing through it; such a chord is called a *diameter*.

The projection of the centre of the circle is a point having the same property in regard to the ellipse, which is therefore called the *centre* of the ellipse. For let aca' be the projection of ACA'; then $ac : ca' = AC : CA'$; but $AC = CA'$, therefore $ac = ca'$. It follows also that *if any two chords bisect one another, their intersection is the centre.*

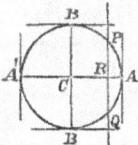 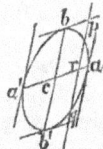

Conjugate Diameters. The tangents at the extremity of a diameter of a circle, AA', are perpendicular to that diameter; if we draw another diameter BB' perpendicular to AA', and therefore parallel to these tangents, the tangents at the extremities of BB' will be perpendicular to BB', and therefore parallel to AA'. It follows that in the ellipse, if we draw a diameter bb' parallel to the tangents at the ends of aa' the projection of AA', this line bb' will be the projection of BB', for parallel lines project into parallel lines; therefore also the tangents at the extremities of bb' will be parallel to aa'. Such diameters are called *conjugate diameters;* they are projections of *perpendicular diameters* of the circle.

Each of the diameters AA', BB' bisects all chords parallel to the other; thus AA' bisects PQ in the point R. Now PQ is projected into a chord pq parallel to bb', and the middle point R is projected into the middle point r. Hence also in the ellipse, *each of two conjugate diameters bisects all chords parallel to the other.*

The assumption here made, that a tangent to the circle projects into a tangent to the ellipse, may be justified as follows. If we take a line PQ cutting the circle in two points, and move it away from the centre until these two points coalesce into one, as at A, the line becomes a tangent. Now when these two points coalesce, their projections must also coalesce; therefore when the line becomes a tangent to the circle, its projection also becomes a tangent to the ellipse.

Relation between ordinate and abscissa. In the circle, if PM, PL be drawn parallel to CB, CA respectively, we know that $CP^2 = CM^2 + MP^2$, and since $CP = CA = CB$, it follows that

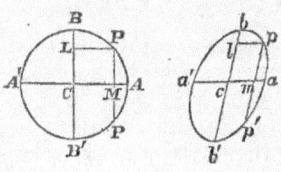

$$\frac{CM^2}{CA^2} + \frac{MP^2}{CB^2} = 1.$$

Hence it is equally true in the ellipse that $\frac{cm^2}{ca^2} + \frac{mp^2}{cb^2} = 1$.

For the ratio of parallel lines being unaltered by parallel projection, $cm : ca = CM : CA$, and $mp : cb = MP : CB$. The line mp is called an *ordinate* or standing-up line, and cm is called an *abscissa* or part cut off. If we write x for cm, y for mp, a for ca, b for cb, the equation becomes

$$\frac{x^2}{a^2} + \frac{y^2}{b^2} = 1.$$

The same relation may be expressed in another form which is sometimes more useful. Namely, observing that the rectangle $a'm \cdot ma = (ca + cm)(ca - cm) = ca^2 - cm^2$, we find that $mp^2 : cb^2 = a'm \cdot ma : ca^2$. This may also be proved directly by observing that it is true for the circle and that the ratios involved are ratios of parallel lines.

This relation shews that when two conjugate diameters are given in magnitude and position, the ellipse is completely determined. For through every point m in aa' we can draw a line parallel to bb', and the points p, p' where this line meets the ellipse are fixed by the equation

$$mp^2 \text{ (or } mp'^2) : cb^2 = a'm \cdot ma : ca^2.$$

Axes. The longest and shortest diameters of an ellipse are conjugate and perpendicular to each other. We may shew, in general that if the distance of a curve from a fixed point o increases up to a point a and then decreases,

the tangent at a, if any, will be
perpendicular to oa. We say *if any*,
because the curve might have a sharp
point at a, and then there would
be properly speaking no tangent *at
a*. Since the distance from o increases up to a and then decreases,
we can find two points p, q, one on
each side of a, such that the lengths
op, oq are equal. Then the perpen-

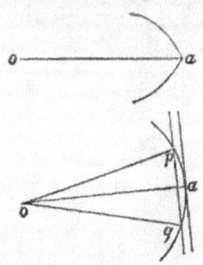

dicular from o on the line pq will fall midway between
p and q. Now suppose p and q to move up towards a,
keeping always the lengths op, oq equal; then the foot
of the perpendicular on pq will always lie between p
and q. When therefore the line pq moves on until p
and q coalesce at a, the foot of the perpendicular will
coalesce with them, or oa is perpendicular to the tangent
at a.

The length oa is called a *maximum value* of the distance from o. It need not be ab-
solutely the *greatest* value, but it
must be greater than the values
immediately close to it on either
side. A similar demonstration applies to a point where the distance,
after decreasing, begins to increase;
that is, to a *minimum* value of the
distance.

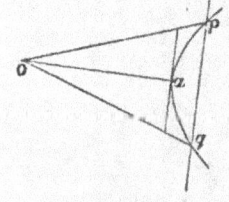

Applying these results to the
Ellipse, we see that the tangents
at the extremities of the longest
and shortest diameters (which of
course must be points of greatest
and least distance from the centre)

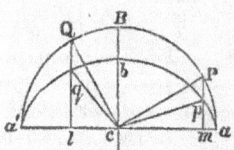

are perpendicular to those diameters. Let bb' be the
shortest diameter, and draw aa' perpendicular to it, and
therefore parallel to the tangents at b, b'; then aa' is
conjugate to bb', and consequently the tangents at a, a'
are parallel to bb', and therefore perpendicular to aa'.

Therefore aa' and bb' are conjugate diameters perpendicular to each other. Now describe on aa' as diameter a circle, $aBa'B'$. If this circle be tilted round the line aa', until B is vertically over b, and then orthogonally projected on the plane of the ellipse, the projection will be an ellipse having aa' and bb' for conjugate diameters, which must therefore be the same as the given ellipse. Hence if Ppm be a line parallel to bb' meeting the circle, ellipse, and aa' in P, p, m respectively, we must have $mp : mP = cb : cB$. Hence the ellipse lies entirely within the circle, and therefore no other distance from the centre is so great as ca or ca'; that is, aa' is the greatest diameter.

The diameters aa' and bb' are called the axes of the ellipse; aa' is the *major* or *transverse* axis, bb' the *minor* or *conjugate* axis. The circle on aa' as diameter is called the *auxiliary circle*.

No other pair of conjugate diameters can be at right angles; for they are projections cp, cq of perpendicular diameters cP, cQ of the circle, and the angle pcq is always greater than the right angle PcQ.

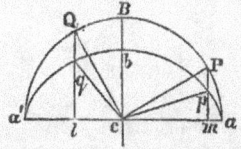

We see, then, that in every case of parallel projection, there are two sets of parallel lines, perpendicular to each other in the original figure, that remain perpendicular to each other in the projected figure.

ELLIPTIC HARMONIC MOTION.

A parallel projection of uniform circular motion is called *elliptic harmonic* motion.

An elliptic harmonic motion may be resolved into two simple harmonic motions of the same period along any two conjugate diameters of the ellipse, these motions differing in phase $\frac{1}{4}$. For we know that the uniform circular motion may be resolved into two such simple harmonic motions along any two perpendicular diameters. And the parallel projection of a simple harmonic motion is clearly another simple harmonic motion, with the same period and phase.

Conversely, *any two simple harmonic motions on different lines, having the same period and differing in phase* $\frac{1}{4}$, *compound into harmonic motion in an ellipse having those two lines for conjugate diameters.* For let the ellipse be constructed; then we have shewn that a circle can be so placed as to have the ellipse for its orthogonal projection. Consequently the two given conjugate diameters are orthogonal projections of two perpendicular diameters of the circle, and the harmonic motions on them are projections of harmonic motions of the same period and phase on the diameters of the circle. But the resultant of these is uniform motion in the circle; therefore the resultant of their projections is the projection of uniform circular motion, namely, harmonic motion in the ellipse.

The *equation* to elliptic harmonic motion is

$$\rho = \alpha \cos (nt + \epsilon) + \beta \sin (nt + \epsilon),$$

where α, β, are two semiconjugate diameters of the ellipse. For the equation to the motion of m is $\rho = \alpha \cos (nt + \epsilon)$ if $\alpha = ca$, and that to the motion of l, having amplitude β, the same period, and phase differing by $\frac{1}{4}$ (and therefore epoch differing by $\frac{1}{2}\pi$, which is a quarter circumference), must be

$$\beta \cos (nt + \epsilon - \tfrac{1}{2}\pi) = \beta \sin (nt + \epsilon).$$

And the motion of ρ is compounded of these two.

The resultant of any number of simple harmonic motions in any directions, having the same period, is elliptic harmonic motion. Let the equations to the different motions be

$$\rho_1 = \alpha_1 \cos (nt + \epsilon_1), \ \rho_2 = \alpha_2 \cos (nt + \epsilon_2), \ldots \rho_n = \alpha_n \cos (nt + \epsilon_n).$$

Expanding these cosines, we have, for example,

$$\rho_1 = \alpha_1 \cos \epsilon_1 . \cos nt - \alpha_1 \sin \epsilon_1 . \sin nt ;$$

and then, adding all together,

$$\rho = \rho_1 + \rho_2 + \ldots + \rho_n = (\alpha_1 \cos \epsilon_1 + \alpha_2 \cos \epsilon_2 + \ldots + \alpha_n \cos \epsilon_n) \cos nt$$
$$- (\alpha_1 \sin \epsilon_1 + \alpha_2 \sin \epsilon_2 + \ldots + \alpha_n \sin \epsilon_n) \sin nt = \alpha \cos nt + \beta \sin nt,$$

if $\alpha = \Sigma \alpha \cos \epsilon$, $-\beta = \Sigma \alpha \sin \epsilon$. This is the equation to elliptic harmonic motion.

It is worth while to notice the meaning of the steps in this demonstration. The expansion

$$\alpha_1 \cos (nt + \epsilon_1) = \alpha_1 \cos \epsilon_1 . \cos nt - \alpha_1 \sin \epsilon_1 . \sin nt$$

is equivalent to a resolution of the simple harmonic motion into two in the same line, differing in phase $\frac{1}{4}$. The epoch of one of these may be assumed arbitrarily, say η; for

$$\alpha_1 \cos (nt + \epsilon_1) = \alpha_1 \cos (nt + \eta + \epsilon_1 - \eta)$$

$$= \alpha_1 \cos (\epsilon_1 - \eta) . \cos (nt + \eta) - \alpha_1 \sin (\epsilon_1 - \eta) . \sin (nt + \eta).$$

This is a particular case of the resolution of elliptic harmonic motion into two simple harmonic constituents, differing in phase $\frac{1}{4}$, the epoch of one being arbitrary (since *any* two conjugate diameters may be chosen). Then the summation $\Sigma \alpha \cos \epsilon . \cos nt = \alpha \cos nt$ means that the resultant of any number of simple harmonic motions of the same period and phase is a simple harmonic motion of that period and phase. Thus all the simple harmonic motions are reduced to two, which differ in phase $\frac{1}{4}$; and the resultant of these, as we know, is elliptic harmonic motion.

COMPOUND HARMONIC MOTION.

If we combine together two simple harmonic motions in different directions with different periods, the resultant motion is periodic if the periods are commensurable, and its period is their least common multiple; if they are incommensurable the path of the moving point never returns into itself so as to form a closed curve. In either case the most convenient way of studying the resultant motion is to convert it into motion on a cylinder, by combining with it a simple harmonic motion perpendicular to its plane which forms a uniform circular motion with one of the components. Suppose, for example, that we wish to study the motion $\rho = \alpha \cos (nt + \epsilon) + \beta \cos mt$ (where α may be taken perpendicular to β) for different values of ϵ. Then we should combine with it a motion $\rho = \gamma \sin (nt + \epsilon)$, where γ is perpendicular to both α and β, and of the same length as α. The two terms $\alpha \cos (nt + \epsilon) + \gamma \sin (nt + \epsilon)$ give a uniform circular motion in a plane perpendicular to β.

Thus we have now to combine a uniform circular motion with a simple harmonic motion perpendicular to its plane. Or, we suppose a generating line to move uniformly round a cylinder, while a point moves up and down it with a simple harmonic motion. This is clearly the same thing as wrapping round the cylinder the curve of positions of the motion $\beta \cos mt$. Hence the path of the motion on the cylinder may always be obtained by wrapping round the cylinder a harmonic curve.

Now the original motion

$$\rho = \alpha \cos (nt + \epsilon) + \beta \cos mt$$

is clearly the projection of this motion on the cylinder upon a plane perpendicular to γ; which plane we may suppose to be drawn through the axis of the cylinder. But by taking different planes through the axis for plane of projection we produce the same effect as by varying ϵ. For this is the same as varying the diameter 2α of the circle on which we project the uniform circular motion $\rho = \alpha \cos (nt + \epsilon) + \gamma (\sin nt + \epsilon)$. And if the same circular motion be projected on two different diameters aa' and bb', the resulting simple harmonic motions will differ in epoch by the angle aob.

We may illustrate this by the case $m = 2n$, when the motion is $\rho = \alpha \cos (nt + \epsilon) + \beta \cos 2nt$. The case $\epsilon = 0$ is always one of special simplicity, being (like the simple harmonic motion) a case of oscillation on a finite portion of a curve.

Let then $om = \beta \cos 2nt = ob \cos 2nt$,

$$mp = \alpha \cos nt = oc \cos nt.$$

Then $am = \beta (1 + \cos 2nt)$

$$= 2\beta \cos^2 nt = ab . \cos^2 nt,$$

but $mp^2 = oc^2 . \cos^2 nt.$

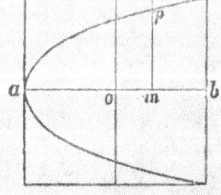

Therefore $mp^2 : oc^2 = am : ab$, or mp^2 varies as am. The curve in which this is the case is called a *parabola;* its two branches extend indefinitely to the right, but only a finite portion of it is traversed by the harmonic motion.

This finite curve, then, is the orthogonal projection of a curve on a cylinder; the axis of which may be (1) vertical, (2) horizontal. In case (1), the curve is made by wrapping round the cylinder a harmonic curve one wave of which will go twice round; or, which is the same thing, by bending into a cylindrical form the spindle-shaped figure here drawn. In case (2), we must wrap round a harmonic curve, *two* of whose waves will go *once* round; the result is something like an ellipse whose plane is bent. The figure obtained by looking at the first along ab or the second along a direction making

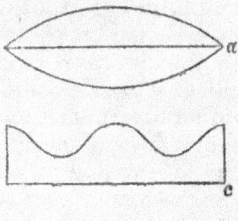

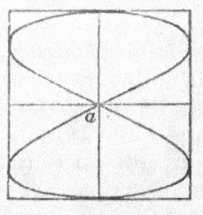

an angle of 45° with co in a vertical plane is here given. A series of intermediate forms is given in Thomson and Tait's *Natural Philosophy*, p. 48. The equation to the figure-of-8 motion is $\rho = \alpha \cos (nt + \frac{1}{4}\pi) + \beta \cos 2nt$, or $\rho = \alpha \cos nt + \beta \cos (2nt + \frac{1}{2}\pi)$. All the intermediate forms can be got by looking at the curve on the cylinder from a sufficient distance and turning it round the axis of the cylinder. For this purpose the curve should either be made in stiff wire or drawn on a glass tube.

Whenever two simple harmonic motions in rectangular directions with commensurable periods are compounded together, there is a certain relation of the phases (viz. the equation is $\rho = \alpha \cos nt + \beta \cos mt$) for which the path resembles that on a parabola in the case just considered; namely, the path is a finite portion of a geometrical curve on which the moving point oscillates backwards and forwards. There is also a path which resembles the figure-of-8 in being symmetrical in regard to each of the perpendicular lines α, β. From a knowledge of these two the intermediate forms may be easily inferred.

The general shape of these two forms may be obtained by a very simple process, which will be understood from a particular example of it. The figures on the opposite

page represent the symmetrical curve and the curve of
oscillation for the case $m : n = 3 : 4$.   Suppose the
amplitudes of the two motions equal, so that the path
is included in a certain square; and draw circles equal
to the inscribed circle of the square touching two of its
sides at their middle points.   Then if one point goes
uniformly three times round $ABCD$ while another goes
four times round FGH, a horizontal line through the
first point and a vertical line through the second will
intersect on the curve which is to be drawn.   The contacts of the curve with the sides of the square are
projections on them of the four points $ABCD$, forming
an inscribed square, and the three points FGH, forming
an equilateral triangle.   For the symmetrical curve these
must be so disposed on their respective circles that their
projections $abcd$ and fgh shall be all distinct and symmetrically placed in regard to the sides of the square.
For the curve of oscillation they must be so placed that
the projections either coincide two and two or are at
the corners of the square.   When the contacts are determined, we must begin at any corner of the square—
say the left-hand bottom corner, Fig. 1, and join the
nearest points cf by a piece of curve convex to the corner;
then the next two, dh, in a similar manner; then bg;
then ac', but as these points are on opposite sides of the
square, the curve has a point of inflexion between them.
The symmetry of the curve will now enable us to complete the figure; or we may apply the same process,
beginning at the adjacent corner gc'.

The curve of oscillation has always to go through a
corner of the square.   If we fix upon the corner d, Fig. 2,
this determines the position of $A'B'C'D'$ and of FGH,
shewing that the curve also goes through the corner b,
but not through either of the other corners.   This determines the position of $ABCD$, because their projections
are thus obliged to coincide two and two.   The motion
takes place on a portion only of the geometrical curve*,
whose continuations, indicated by the dotted lines, re-
semble a parabola in shape.

* On the equations and geometrical character of these curves, see
Braun, "Ueber Lissajous' Curven," *Math. Annalen*, VIII. p. 567.

Fig. 1.

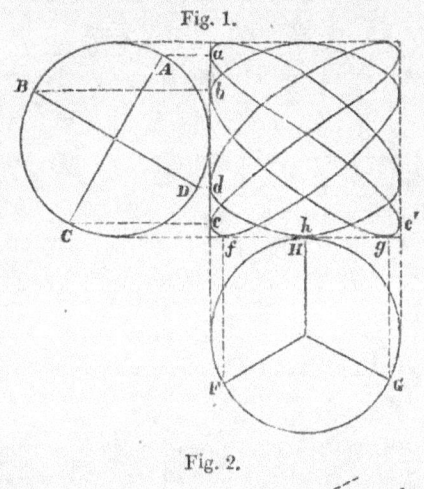

Fig. 2.

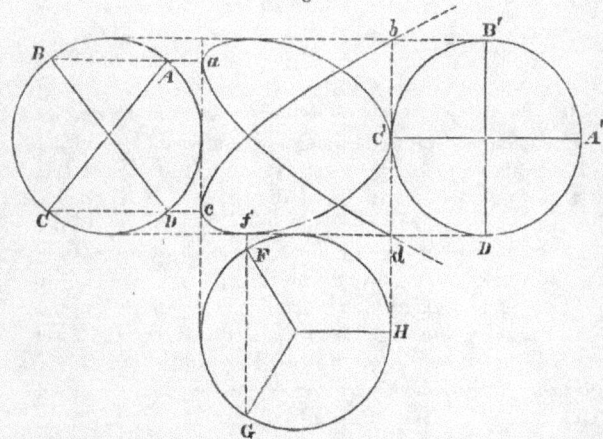

Examples of harmonic motion, simple and compound, occur in the vibrations of elastic solids, the rise and fall of the tides, the motion of air-particles when transmitting sound, and of the ether in carrying radiations of light and heat. The important theorem of Fourier, that every motion which exactly repeats itself after a certain interval of time is a compound of harmonic motions, will be proved in the Appendix.

PARABOLIC MOTION.

It was observed by Galileo that if a body be let slide down a smooth inclined plane, the lengths passed over from the beginning of the motion are proportional to the squares of the times. That is, if the body goes a length a in the first second, then in the first t seconds it will go a length at^2. It is easy to see that the lengths passed over in successive seconds are proportional to the successive odd numbers; for in the nth second the distance travelled is

$$an^2 - a(n-1)^2 = a(2n-1).$$

Now $2n-1$ is the nth odd number.

In the extreme case included in this law, when a body falls vertically *in vacuo*, it falls in the first second 490·4 centimeters, or 16·1 feet.

To determine the curve of positions for this motion we must compound it with a uniform rectilinear motion in a direction at right angles to it. More generally, if we combine with it a uniform rectilinear motion in any direction (this is what happens with a body thrown obliquely upwards) we obtain a motion which is called *parabolic*, from *parabola*, the name of the curve described.

The equation to the rectilinear motion of a body falling vertically or down an inclined plane is evidently $\rho = \alpha + t^2\gamma$, where α is the step to the starting point, and γ is the step taken by the body in the first second. Combining with this the uniform rectilinear motion $\rho = t\beta$, we find for the equation to the parabolic motion $\rho = \alpha + t\beta + t^2\gamma$.

Let $ac = \gamma$, $ab = \beta$, $am = t^2\gamma = t^2 . ac$, $mp = t\beta = t . ab$; then p is the position of the moving point after t seconds. And $am : ac = t^2 : 1 = mp^2 : ab^2$; or am varies as mp^2. We have already met with this property in a curve described by compound harmonic motion. We shall now prove that this curve (the parabola) is a central projection of a circle; in fact, it is *the shadow of a circle cast on a horizontal plane by a luminous point on a level with the highest point of the circle.*

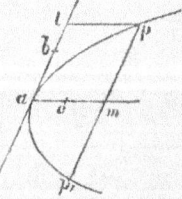

Let aqb be the circle, of which b is the highest point, and a the lowest is resting on the horizontal plane ; v the centre of projection. From q any point on the circle draw qn perpendicular to ab; let am be the projection of an, pm of qn, and ap of the circular arc aq. The line pm will not in general be perpendicular to am, but will be parallel to the common tangent at a to the two curves. Then by similar triangles, we find

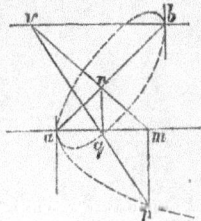

$$pm : qn = vm : vn = ab : nb.$$

Now $\qquad qn^2 = an \cdot nb$;
therefore

$$pm^2 : an \cdot nb = ab^2 : nb^2, \text{ or } pm^2 : ab^2 = an : nb = am : vb.$$

Thus pm^2 varies as am, which was to be proved.

It is clear that $pm = mp'$, because their squares are both proportional to am; thus the line am bisects all chords parallel to the tangent at a. Because it has this property in common with the diameter of an ellipse or a circle, it is called a *diameter* of the parabola. We shall now shew that a diameter may be drawn through every point of a parabola, and that all these diameters are parallel.

Suppose that in the motion $\rho = \alpha + t\beta + t^2\gamma$, the epoch from which the time is reckoned is changed to τ seconds later. Then t, the number of seconds from the old epoch to any instant, is greater by τ than t', the number of seconds from the new epoch to that instant, or $t = t' + \tau$. Therefore

$$\rho = \alpha + (t' + \tau)\beta + (t' + \tau)^2\gamma = \alpha + \tau\beta + \tau^2\gamma + t'(\beta + 2\tau\gamma) + t'^2\gamma.$$

Hence the equation is of the same form as before, except that α and β are changed. The new α is of course the step oa'; the step aa' being clearly $\tau\beta + \tau^2\gamma$. The new β is $\beta + 2\tau\gamma$; thus we must draw $a'k$ equal and parallel to ab, and then $kb' = 2\tau \cdot ac$. The direction of $a'b'$ is most easily determined by producing it backwards to t.   Then

$$tm : kb' = ma' : ab = \tau : 1.$$

Therefore $tm = \tau \cdot kb' = 2\tau^2 \cdot ac = 2am.$

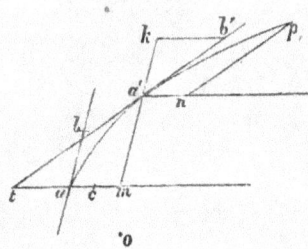

Now the lines $a'b'$ and $a'n$ are tangent and diameter of the parabola, having the same properties as ab and am.

For $np^2 : a'b'^2 = t'^2 : 1 = a'n : ac,$

so that np^2 varies as $a'n$.   Hence we derive the following properties of the parabola:—

1. If from any point a' we draw $a'm$ parallel to the tangent at a to meet the diameter through a, and make $at = ma$, then $a't$ is the tangent to the parabola at a'.

2. If pn be drawn parallel to the tangent at *any* point a' of a parabola to meet the diameter through its point of contact at n, pn^2 is always proportional to $a'n$.

It is easy to shew that *one* of the diameters is perpendicular to the tangent at its vertex.   For we can so determine τ (positive or negative) that $\beta + 2\tau\gamma$ shall be perpendicular to γ.   We have only to draw al perpendicular, bl parallel, to ac, and then make $\tau = bl : ac$. This diameter is called the *axis*, and the parabola 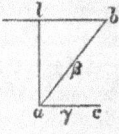 is symmetrical on either side of it.   Thus of the parabola drawn on p. 34, the axis is ab.   A stone thrown into the air describes approximately a parabola whose axis is vertical.

CHAPTER II. VELOCITIES.

THE DIRECTION OF MOTION. (TANGENTS.)

EUCLID defines the tangent to a circle as a line which meets it but does not cut it; and he shews that it is always perpendicular to the radius through the point of contact. This line may also be regarded as the final position of a chord which moves parallel to itself until its two points of intersection with the circle coalesce into one point. As the foot of the perpendicular from the centre is always midway between these two points, it must coincide with them when they coincide together.

We may find the tangent to an ellipse or parabola at any point by means of the remark already made, that the projection of the tangent to a curve is tangent to the

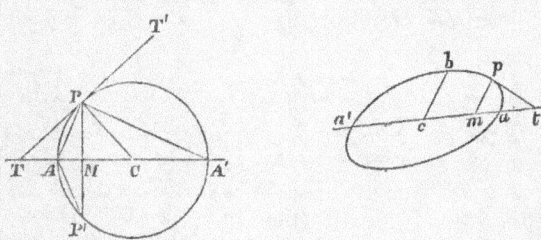

projection of the curve. Since, in the circle, the triangles CMP, CPT are similar, where PM is drawn perpendicular to CT; we have $CM : CP = CP : CT$, or $CM . CT = CA^2$. It follows that in the ellipse also, if we draw the tangent pt to meet the diameter aa', and pm parallel to the conjugate diameter cb, we must have $cm . ct = ca^2$. Moreover, in the circle $TA : AM = TA' : MA'$. For, producing PM to meet the circle again at P', we know that the angle $TPA = AP'P = APM$, or PA bisects the angle TPM. So also PA' bisects the external angle MPT'. Therefore

$$TA : AM = TP : PM = TA' : MA'.$$

This also then is true for the ellipse;

$$ta : am = ta' : ma' \text{ or } am : ma' = at : a't.$$

Thus each of the two segments aa', mt, divides the other internally and externally in the same ratio. Four points so situated are called a *harmonic range*, because the lengths ta, tm, ta' are in harmonic progression.

The central projection of a harmonic range is also a harmonic range. Let $abcd$, four points on a straight line, be central projections of $ABCD$ from a point v. Twice the area of the triangle avb is equal to the rectangle $ab.vm$ and also to $va.vb.\sin avb$.

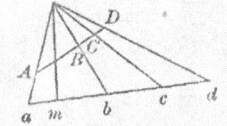

So also $cd.vm = vc.vd \sin cvd$.

Hence

$$ab.cd.vm^2 = \sin avb.\sin cvd.va.vb.vc.vd.$$

Similarly

$$ac.bd.vm^2 = \sin avc.\sin bvd.va.vb.vc.vd.$$

Therefore

$$ab.cd : ac.bd = \sin avb.\sin cvd : \sin avc.\sin bvd.$$

The ratio $ab.cd : ac.bd$ is called a *cross-ratio* of the four points $abcd$ (ratio of the ratios in which ad is divided by b and c). And $\sin avb.\sin cvd : \sin avc.\sin bvd$ is called a cross-ratio of the four lines va, vb, vc, vd. Hence as the cross-ratio of the points $abcd$ is equal to the corresponding cross-ratio of the lines va, vb, vc, vd, it is also equal to the corresponding cross-ratio of the points $ABCD$. Thus *a cross-ratio is unaltered by central projection*. But the four points are harmonic when $AB.CD = AD.BC$ or when $AB.CD : AD.BC = 1$. In this case $ab.cd : ad.bc$ is also unity, or $abcd$ form a harmonic range.

Now the parabola being a central projection of a circle, the points T, A, M, A' are projected into t, a, m and a point at an infinite distance, say a'. Since a' is at an infinite distance $ta' : ma' = 1$. But since the four points form a harmonic range, $ta : am = ta' : ma' = 1$, or $ta = am$, as we before proved by another method. This determines the tangent to a parabola at any point p.

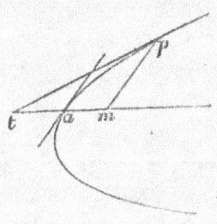

To determine the tangent to the harmonic curve, we must remember that it is formed by unrolling an ellipse from a cylinder. Let ac be the ellipse, ab its orthogonal projection, the circular section of the cylinder by a plane

Fig. 1.   Fig. 2.   Fig. 3.

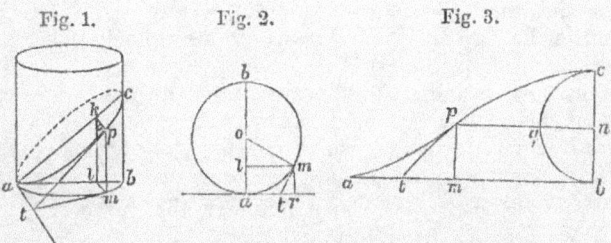

perpendicular to the axis, pm perpendicular to that plane, pt and mt tangents to the ellipse and circle respectively, at tangent to both of them. A plane touching the cylinder along the line pm will clearly cut the planes of the ellipse and circle in tangents to them at p and m, which must meet at t on at the line of intersection of those planes. The second figure represents the circle ab in the plane of the paper, and the third figure the result of unwrapping one-half the ellipse. Now

$$mp : bc \text{ (fig. 1)} = lk : bc = al : ab \text{ (fig. 1 or 2).}$$

Therefore (since $mp = bn$, fig. 3)

$$bn : bc = al : ab \text{ (fig. 2),}$$

and consequently

$$2qn : bc = 2lm : ab = lm : om = rm : tm = al : tm.$$

That is $al : mt = 2nq : bc$, but $pm : al = bc : ab$, therefore

$$pm : mt = 2nq : ab \text{ (of fig. 2)} = \pi nq : ab \text{ (of fig. 3)},$$
since $\qquad ab \text{ (fig. 3)} = \tfrac{1}{2}\pi . ab \text{ (fig. 2)}.$
Thus $\qquad pm : mt = \pi . nq : ab.$

Hence the inclination of the tangent is greatest when
nq is greatest, or when n is the centre of bc. The point
of greatest inclination is called a *point of inflexion*, because the curve stops bending upwards and begins to bend
downwards.

EXACT DEFINITION OF TANGENT.

We have regarded the tangent at a point a of a curve
as the final position of a line cutting the
curve in two points p, q, when the line is
made to move so that p, q coalesce at a.
This method indeed will always find the tangent *when there is one.* But we have seen
that when the curve has a sharp point at a
there is properly no tangent *at* the point a.

In the case of a sharp point in the curve, we may
draw a line ab through it, and then
turn this line round until b moves
up and coalesces with a. The final
position at of this line may be
called the tangent *up to a.* Similarly if we draw a line ac cutting
the curve on the other side, and
turn this round until c moves up
to a, the final position at' of this

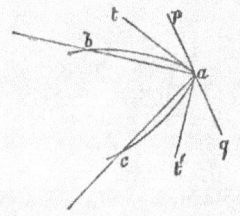

line may be called the tangent *on from a.* So that
we have a tangent *up to a* and a tangent *on from a,* but
no tangent *at a,* properly so called.

The final position of ab when b has moved up to a, is
however not so well defined in this case as when there
is no sharp point. For then if we turn the line a little
too far round a, it will cut the curve on the other side.
But when there is a sharp point, there are intermediate

positions between *at* and *at'*, such as *pq*, in which the line does not cut the curve on either side of *a*.

To improve this definition, we observe that the true tangent *at* has the property that if we turn it ever so little in the direction *ab* it will cut the curve between *a* and *b*. Hence, when such a line *ab* has been drawn, it is always possible to find a point *x* on the curve such that *ax* shall lie between *at* and 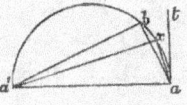 *ab*; or (which is the same thing) so that *ax* shall make with *at* an angle less than that which *ab* makes. This is very obvious in the case of the circle, for the angle *bat* = *ba'a*, and *xat* = *xa'a*; so that we have only to draw through *a'* a line *a'x* making with *a'a* a less angle than *a'b* makes.

This rule may be stated as follows: If *at* is the tangent at the point *a*, it is possible to find a point *x* on the curve near to *a* so that the angle *xat* shall be less than any proposed angle, however small. For let *tab* be the proposed angle; however small it is, the line *ab* must cut the curve if *at* is a tangent, and then we have only to take a point *x* between *a* and *b*.

The proof fails, however, when the curve is wavy. In the figure we can take a point *x* between *a* and *b* so that *ax* does  not lie between *at* and *ab*. This only means that we have begun too far away from *a*. If we take *b* somewhere between *a* and the nearest point of inflexion *c*, the proof of the rule will hold good.

So guarded, the rule amounts to Newton's criterion for a tangent. Even this criterion, however, is baffled by some curves which can be conceived. Suppose two circles to touch one another at the point *a*, and between these circles let us draw a wavy curve, like the harmonic curve, except that the waves become smaller and smaller as they approach *a*; and let us suppose the *shape* of the waves, that is, the ratio of their height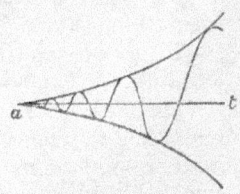

to their length, to be kept always the same. Then it
will be impossible to take b so near to a that there shall
be no point of inflexion between them. Also it is clear in
this case that there is *no* real tangent at a; for however
near we get to a, the direction of the curve sways from
side to side through the same range.

If, however, the waves are so drawn that the ratio
of their height to their length
becomes smaller and smaller as
they approach a, so that they
get more and more flat without
any limit, then although the
proof of the rule fails as before,
there *is* a real tangent at a,
namely, the common tangent at
to the two circles.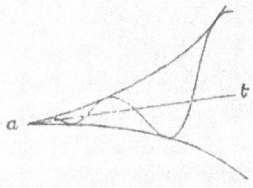

In both these cases our criterion for a tangent is
satisfied; that is to say, there is a line at such that by
taking a point x on the curve near enough to a, the
angle xat can be made less than any proposed angle. Yet
in one of these cases this line at is a tangent, and in
the other it is not. We must therefore find a better
criterion, which will distinguish between these cases.

The tangent to a circle has the following property.
If we take *any* two points p and q between a and b, the
chord pq makes with the tangent
at a an angle less than aob. For
the angle between pq and at is
equal to aom, where om is perpendicular to pq. Let pq be called
a chord inside ab, even if p is
at a or q is at b. Then we can
find a point b such that every chord inside ab makes
with at an angle less than a proposed angle, however
small. For we have only to draw the angle aob a little
less than the proposed angle.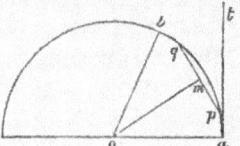

Now the second of our exceptional curves, that which
really has a tangent, has also this property, that we can
find a point b so near to a that every chord inside ab
shall make with at an angle less than a proposed angle,

however small. For since the waves get flatter and flatter without limit, the tangents at the successive points of inflexion make with *at* angles which decrease without limit. We have then only to find a point of inflexion whose tangent makes with *at* an angle less than the proposed angle, and take *b* at this point or between it and *a*.

But the first curve has not this property, for the inclinations of the tangents at the points of inflexion are always the same, and any one of these counts as a chord inside *ab*.

We shall now therefore make this definition :

When there is a line *at* through a point *a* of a curve having the property that, any angle being proposed, however small, it is always possible to find a point *b* so near to *a* on one side that every chord inside *ab* makes with *at* an angle less than the proposed angle; then this line *at* is called the tangent of the curve up to the point *a* on that side.

When there are tangents up to the point *a* on both sides, and these two are in one straight line, that straight line is called the tangent *at a*. In this case the curve is said to be *elementally straight* in the neighbourhood of *a*. It has the property that the more it is magnified, the straighter it looks.

Going back to our first and simpler definition of a tangent, as the final position of a line *pq* which is made to move so that *p* and *q* coalesce at *a*, we see that not only does it always find the tangent when there is one, but that when there is not, the final position of *pq* will not be determinate, but will depend upon the way in which *p* and *q* are made to coalesce at *a*. When therefore this method gives us a *determinate* line, we may be sure that that line is really a tangent.

VELOCITY. UNIFORM.

The problem which we have now to consider is the following :—Suppose that we know the position of a point at every instant of time during a certain period, it is required to find out *how fast it is going* at every instant during the same period. For example, in the simple harmonic motion described in the last chapter, we know the position of the point m by geometrical construction; namely we determine the position of p by measuring an arc ap on the circle proportional to the time, and then we draw a perpendicular pm to the diameter aa'. The problem is to find out how fast the point m is moving when it is in any given position. The rate at which it is moving is called its *velocity*.

Let us now endeavour to form a clearer conception of this quantity that we have to measure; and for this purpose let us consider the simplest case, that of uniform motion in a straight line. We say of a train, or a ship, or a man walking, that they go at so many *miles per hour*; of sound, that it goes 1090 *feet per second*; of light, that it goes 200,000 *miles per second*. These statements seem at first to mean only that a certain space has been passed over in a certain time; that the man, for instance, has in a given hour walked so many miles. But because we know that the motion is uniform, we are hereby told not only how far the man walks in an hour, but also how far he walks in any other period of time. In walking on a French road, for example, it is convenient to walk about six kilometers per hour. Now this is one kilometer per ten minutes, and that is the same thing as one hectometer per minute. In what sense *the same thing?* It is not the same thing to walk a hectometer in one minute as it is to walk six kilometers in one hour. But *the rate at which one is moving* is the same during the minute as it is during the hour. Thus we see that to say how fast a body is going is to make a statement about its state of motion at any instant and not about its change of position in any length of time. The velocity of a moving body is an instantaneous property of it which may or may not change from instant to instant; and the peculiarity of *uniform* motion, in which equal spaces are traversed in equal times, is that the velocity remains constant throughout the motion; a body which moves uniformly is always going at the same rate.

But although this rate is a property of the motion which belongs to it at a given instant, we cannot measure it instantaneously. In order to find out how fast you are walking at a particular instant, you must keep on walking at that same rate for a definite time, and then see how far you have gone. Only, as we noticed before, it does not matter what that definite time is. Whether you find that you have walked one hectometer in a minute, or one kilometer in ten minutes, or six kilometers in an hour, the velocity so measured is the *same* velocity. Now for comparing velocities together, it is found convenient to refer them all to the same interval of time. Which goes faster, sound at 1090 feet per second, or a molecule of oxygen in the air at seventeen miles a minute? Clearly we must find how far the molecule of oxygen would go in a second, and compare that distance with 1090 feet. For scientific purposes the *second* is the period of time adopted in measuring velocities; and we may say that we know the rate at which a thing is moving when we know how far it would go in a second if it went at that same rate during the second.

A velocity, then, is measured by a certain *length;* namely, the distance which a body having the velocity during a second would pass over in that second. It may therefore be specified either graphically, by drawing a line to represent that length on a given scale, or by numerical approximation. When a velocity is described as so many *centimeters per second* it is said to be expressed in *absolute measure.* Thus the *absolute unit* of velocity is one centimeter per second. The absolute measure of six kilometers per hour is 166⅔. More generally we may say that the unit of velocity is one unit of length per unit of time.

This last statement is sometimes expressed in another way. Let $[V]$ denote the unit of velocity, $[L]$ the unit of length, and $[T]$ the unit of time; then $[V] = \dfrac{[L]}{[T]}$. Here the word *per* has been replaced by the sign for *divided by:* now it is nonsense to say that a unit of velocity is a unit of length *divided by* a unit of time in the ordinary sense of the words. But we find it convenient to give a new meaning to the words "divided by," and to the symbol which shortly expresses them, so that they may be used to mean what is meant by the word *per* in the expression "miles per hour." This convenience is made manifest when we have to change from one unit to another. Suppose, for instance, that we want to compare the unit of velocity one centimeter per second with another unit, one kilometer per hour. We shall have

$$\text{first unit} = \frac{\text{one centimeter}}{\text{one second}},$$

$$\text{second unit} = \frac{\text{one kilometer}}{\text{one hour}},$$

consequently

$$\frac{\text{second unit}}{\text{first unit}} = \frac{\text{one kilometer}}{\text{one centimeter}} \cdot \frac{\text{one second}}{\text{one hour}} = \frac{100,000}{3600}$$
$$= 27\cdot7.$$

We might have got to the same result by saying that one kilometer per hour is 100,000 centimeters per 3600 seconds, that is, 27·7 centimeters per second. Hence if we give to the symbol of division this new meaning, and then treat it by the rules applicable to the old meaning, we arrive at right results; and we save ourselves the trouble of inventing a new symbol by using the old one in a new sense.

Another way of expressing the equation $[V] = [L]:[T]$ is to say that velocity is a quantity of dimensions 1 in length and − 1 in time.

Velocity is a *directed* quantity; and therefore is not fully specified until its magnitude and direction are both given. The velocity of translation of a rigid body is adequately represented by a straight line of proper length and direction drawn *anywhere*. Consequently it is a *vector* quantity, in the sense already explained.

In the uniform rectilinear motion $\rho = \alpha + t\beta$, the step taken in one second is β, which is therefore the velocity. When the step *op* from a fixed point *o* to the moving point p is denoted by ρ, the velocity of p is denoted by $\dot{\rho}$. Thus if

$$\rho = \alpha + t\beta, \text{ then } \dot{\rho} = \beta.$$

In the uniform circular motion

$$\rho = ai \cos(nt + \epsilon) + aj \sin(nt + \epsilon),$$

the distance travelled in one second is na, and the direction of the motion is pt. Hence if oq be drawn parallel to pt, the velocity is represented by $n \cdot oq$. Now oq is what op will become after a quarter period; that is, after the angle $nt + \epsilon$ has been increased by a right angle. Thus

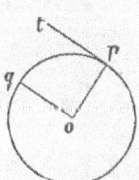

$$oq = ai \cos(nt + \epsilon + \tfrac{1}{2}\pi) + aj \sin(nt + \epsilon + \tfrac{1}{2}\pi);$$

and

$$\rho = n \cdot oq = nai \cos(nt + \epsilon + \tfrac{1}{2}\pi) + naj \sin(nt + \epsilon + \tfrac{1}{2}\pi).$$

The rule to find $\dot{\rho}$ from ρ is therefore, in the case of uniform circular motion: multiply by n, and increase the argument by $\tfrac{1}{2}\pi$.

VELOCITY.   VARIABLE.

To make more precise the idea of a velocity which varies continuously with the time, let us consider the case of two parallel lines of rail, on one of which a train starts from rest and gradually increases in speed up to twenty miles an hour, while on the other a train runs uniformly in the same direction at 10 miles an hour. We will suppose the second train to be so long that a traveller in the first train has always some part of it immediately opposite to him. At starting, the uniform train will appear to this traveller to be gaining on him at the rate of 10 miles an hour; but as his own train gets up speed, this rate of gaining will diminish. At the end, when the variable train is going 20 miles an hour, the uniform train will be *losing* 10 miles an hour. There must have been some moment between these two states of things at which the uniform train was seen to stop gaining and to begin

4—2

to lose. At that moment the variable train was going 10 miles an hour.

In the same way if we suppose the uniform train to go at any other velocity less than 20 miles an hour, there will be an instant at which it will appear to a traveller in the variable train to stop gaining and to begin to lose. This will be the instant at which the variable train having hitherto been travelling at a less velocity, just acquires the velocity of the uniform train, and then, acquiring a still greater velocity, proceeds to gain upon it.

When then we say that at a certain instant a train is going v miles an hour, we mean that a train moving uniformly v miles an hour on a parallel line of rails would appear from the first train to stop. If the velocity of the variable train is continually increasing or continually decreasing, the uniform train will appear to reverse its motion; but if the velocity after increasing up to that point began to decrease, or after decreasing began to increase, the uniform train would seem to stop momentarily and then go on in the same direction.

By these considerations we have reduced the case of an instantaneous velocity of any magnitude to the case of stoppage or zero velocity, which can be readily observed and conceived.

In the motion of a falling body, for example, we have $s = t^2 a$, where a is the distance fallen in the first second, and s the distance fallen in the first t seconds. Suppose another body to move uniformly downwards with velocity v; in t seconds it will have passed over a distance $s_1 = vt$. Thus the distance between the two bodies is $s_1 - s = vt - at^2$. Therefore $4a(s_1 - s) = 4avt - 4a^2t^2 = v^2 - (v - 2at)^2$. This quantity continually increases so long as $v - 2at$ diminishes by the increase of t, that is, until $v = 2at$; then it begins to diminish again. Hence at the moment when $v = 2at$, or $t = v : 2a$, the uniformly moving body stops gaining on the falling body and begins to lose. Consequently the velocity of the falling body at the time t is $2at$. Or if $s = at^2$, then $\dot{s} = 2at$.

It appears therefore that a body falling freely in vacuo has at the end of the first second a velocity of 32·2 feet per second, at the end of the next twice that velocity, and so on ; a velocity of 32·2 feet per second being added to it in every second.

The problem of finding the velocity of a moving point is the same as that of drawing a tangent to its curve of positions. Let qpq' be the curve of positions of a moving point k, and let rpr' be the tangent to it at p. 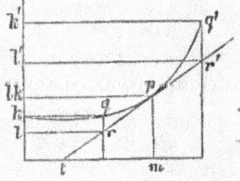 Let a point l move so as to have this tangent for its curve of positions ; then since it moves over a distance equal to mp in a time represented by tm, its velocity is represented by $mp : tm$. Now as the figure is drawn, the curve being wholly on the upper side of the tangent, the point k is always above l, because q is above r ; but if we suppose the line qr to move parallel to itself horizontally, r will gain upon q until it comes up to it at p, and then q will gain upon r. Therefore the same is true of their orthogonal projections ; l will gain upon k, until it comes up with it at the point lk ; 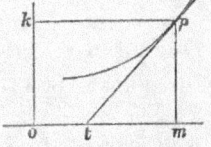 after that it will lose. At this point, consequently, the velocity of k is equal to that of l ; or it is $mp : tm$. Hence this rule : to find the velocity of the point k in any position, draw the tangent pt at the corresponding point of the curve of positions ; then, pm being parallel to ok, the velocity is $mp : tm$.

Now the tangent at any point p of the curve of positions, if there is one, is found by moving the chord joining two points of the curve until these two points coalesce at p. We shall now shew how to translate this rule so as to derive from it a method of finding the velocity without drawing the curve of positions.

The *mean* velocity of a moving point during any interval is that uniform velocity with which the point would make the same step during the interval that it has ac-

tually made. If a body gets from a
to b, by any path, with any variation
of speed, in t seconds, its mean velocity during the interval is $ab : t$. The
direction of it is the straight line ab.

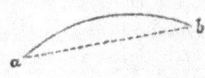

When the motion
is rectilinear, the mean velocity is of course simply the
distance traversed divided by the time of travelling.

Now let k, l be two positions of a moving point, q,
r the corresponding points of the
curve of positions, qm, rn vertical.
Then mn represents the time of
travelling from k to l; and consequently the mean velocity during
this interval is represented by kl :

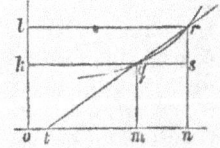

mn. But this is the uniform velo-
city whose curve of positions is the chord qr. Hence *the
chord joining two points on the curve of positions is itself
the curve of positions of the mean velocity during the corresponding interval.* If the chord cuts the line omn in t
the mean velocity is $mq : tm$.

The tangent at a point p is obtained by moving the
chord qr till its ends coalesce at p; and the tangent is
curve of positions of the instantaneous velocity corresponding to the point p. Hence *the instantaneous velocity at
any instant (when there is one) may be obtained from the
mean velocity of an interval by making both ends of the
interval coincide with that instant.* This appears to be
nonsense, because there is no interval when the two ends
coincide. But an example will shew what is the meaning
of the rule.

Let us take again the case of a falling body, $s = at^2$;
it is required to find the instantaneous velocity at the
time t. In the interval between t_1 seconds and t_2 seconds
after the beginning of the time, the distance travelled is
$s_2 - s_1 = at_2^2 - at_1^2$; therefore

$$\text{mean velocity} = \frac{s_2 - s_1}{t_2 - t_1} = a\,\frac{t_2^2 - t_1^2}{t_2 - t_1} = a\,(t_1 + t_2).$$

Thus the mean velocity in the interval from t_1 to t_2 is

$a\,(t_1 + t_2)$. This becomes the instantaneous velocity at the time t when both t_1 and t_2 coincide with t. But then it becomes $2at$, the value we have already found for it. And this result is clearly independent of the *way* in which t_1 and t_2 approach the value t.

To extend this result, suppose the distance travelled to vary as the nth power of the time, where n is an integer; that is, $s = at^n$. Then we shall have in the interval $t_2 - t_1$

$$\text{mean velocity} = \frac{s_2 - s_1}{t_2 - t_1} = a\,\frac{t_2^{\,n} - t_1^{\,n}}{t_2 - t_1}$$

$$= a(t_2^{\,n-1} + t_2^{\,n-2}t_1 + t_2^{\,n-3}t_1^{\,2} + \ldots + t_2 t_1^{\,n-2} + t_1^{\,n-1})$$

(the division here made may be readily verified by multiplying the result by $t_2 - t_1$). Now the quantity in brackets consists of n terms, each of which becomes t^{n-1} when we make t_1 and t_2 each equal to t. Thus the instantaneous velocity is nat^{n-1}. Or we may now say that when $s = at^n$, then $\dot{s} = nat^{n-1}$. The rule for getting $\dot{s}$ from s in this case is: multiply by the index of t, and then diminish that index by 1.

Next suppose that n is a commensurable fraction, the quotient of p by q, where p and q are integers. Suppose that z is a quantity such that $z^q = t$, then $t^n = z^{qn} = z^p$ since $p = qn$. Hence we have $s = at^n = az^p$, while $t = z^q$. Consequently in the interval $t_2 - t_1$

$$\text{mean velocity} = \frac{s_2 - s_1}{t_2 - t_1} = a\,\frac{z_2^{\,p} - z_1^{\,p}}{z_2^{\,q} - z_1^{\,q}}$$

$$= a\,\frac{z_2^{\,p-1} + z_2^{\,p-2}z_1 + \ldots + z_1^{\,p-1}}{z_2^{\,q-1} + z_2^{\,q-2}z_1 + \ldots + z_1^{\,q-1}},$$

where the factor $z_2 - z_1$ has been divided out of the numerator and denominator of the fraction. Making z_1 and z_2 both equal to z, we have

$$\text{instantaneous velocity} = a\,\frac{pz^{p-1}}{qz^{q-1}} = naz^{p-q}.$$

Now $z^p = t^n$ and $z^q = t$, therefore $z^{p-q} = t^{n-1}$. The velocity is therefore again $\dot{s} = nat^{n-1}$. It appears therefore that the rule stated above applies equally whether n is an integer or a commensurable fraction.

The proof that the same rule holds good when n is negative is left as an exercise to the reader.

We may now describe shortly the process for finding $\dot{s}$ when s is given in terms of t. In the fraction

$$\frac{s_2 - s_1}{t_2 - t_1},$$

substitute the values of s_2 and s_1 in terms of t_2 and t_1; strike out any common factors from numerator and denominator; then omit the suffixes.

EXACT DEFINITION OF VELOCITY.

The same difficulties occur in regard to velocities that we have already met with in regard to tangents. When a billiard-ball is sent against a cushion and rebounds, its velocity seems to be suddenly changed into one in another direction. If this were so, we could not speak of a velocity *at the instant* of striking; though we might speak of a velocity *up to* that instant and a velocity *on from it*. Such an event would be indicated by a sharp point in the curve of positions, so far as sudden change in the *magnitude* of the velocity is concerned, or by a sharp point in the path of the moving body, in case of sudden change in *direction*. And still greater difficulties may be conceived, when the curve of positions is like the curves on p. 45, 46, with an infinite number of waves.

It is true that there is some reason to believe that sudden changes of velocity never actually occur in nature; that the billiard-ball, for example, compresses the cushion, and while so doing loses velocity at a very rapid rate, yet still not suddenly; and acquires it again as the cushion recovers its form. However, we cannot deal directly with such motions as occur in nature, but only with certain ideal motions, to which they approximate; and in these ideal motions such difficulties may occur. It is therefore necessary to find a criterion for the existence of a velocity at a given instant. In this we shall follow our previous investigation in regard to tangents.

Our first criterion was this: If ta is the tangent up to a point a, it is possible to find a point x on the curve so that the angle xat shall be less than a proposed angle, however small. Suppose the curve to be curve of positions of some rectilinear motion. Take a horizontal line ou, one centimeter

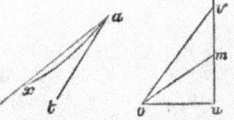

long; draw uv vertical, ov parallel to ta, om parallel to xa. Then the angle mov is equal to xat. Also uv is the instantaneous velocity corresponding to the point a in the curve, and um is the mean velocity in the interval corresponding to xa. If the angle xat or, which is the same thing, the angle mov, can be made less than any proposed angle, it follows that mv can be made less than any proposed length. Therefore, *if uv be the velocity up to a certain instant, it is possible to find an interval ending at that instant in which the mean velocity shall differ from uv less than by a proposed quantity, however small.* That is, by reckoning the mean velocity in a sufficiently small interval, we can make it as close an approximation as we like to the instantaneous velocity.

To define the velocity *on from* an instant, we must take an interval *beginning* at that instant.

The more accurate criterion of the tangent is that x can be so taken that every chord inside ax shall make with at an angle less than a proposed angle. To express the corresponding criterion for a velocity, let us speak of the mean velocity in an interval of time included within a certain interval as a mean velocity inside that certain interval. Then the criterion is that *if v is the velocity up to a certain instant, it is possible to find an* *interval ending at that instant such that every mean velocity inside it shall differ from v less than by a proposed quantity, however small.*

This criterion applies to variable velocity in rectilineal motion in the first instance; but it clearly extends to determination of the *magnitude* of the velocity in curvilinear motion, when that has been represented upon a straight line in the manner used for determining its curve of positions. But we may so state the criterion as to give a direct definition of velocity as a vector in all cases of motion.

Let *ba* be a portion of the path of a moving point, and *p*, *q* two positions either intermediate between *b* and *a*, or coinciding with either of them. Let the mean velocity from *p* to *q* (viz. the step *pq* divided by the time of taking it) be called a mean velocity inside *ba*. Let *ov* represent the velocity up to *a* in magnitude and direction, and *om* the mean velocity in *pq*. Then it is possible to choose *b* so that every mean velocity inside *ba* shall differ

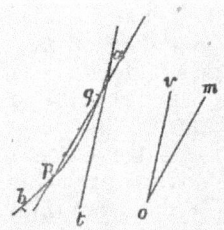

from *ov* less than by a proposed quantity, however small. We say that *om* differs from *ov* by the step *mv*, and it is meant that *mv* is shorter than a proposed length.

When there is a line *ov* having this property, there is said to be a velocity up to the point *a*, and *ov* is that velocity. The velocity *on from a* is defined in a similar manner. When these two are *equivalent* (have the same magnitude and direction) we speak of *ov* as the velocity *at a*. The motion is then said to be *elementally uniform* in the neighbourhood of *a*.

The criterion may be illustrated by applying it to the case $s = at^n$. Let t_0, t_1, t_2, t be four quantities in ascending order of magnitude; we propose to shew that nat^{n-1} is the velocity at the time t. We know that the mean velocity between t_1 and t_2 is

$$a\,(t_1^{n-1} + t_1^{n-2} t_2 + \ldots + t_2^{n-1}).$$

The quantity between brackets consists of n terms, each of which is greater than t_0^{n-1} and less than t^{n-1}. Hence the mean velocity is greater than nat_0^{n-1} and less than nat^{n-1}. The difference between these can evidently be made smaller than any proposed quantity by taking t_0 sufficiently near to t. But the mean velocity from t_1 to t_2 differs still less from nat^{n-1} than nat_0^{n-1} does. Hence it is possible to choose an interval, from t_0 to t, such that the mean velocity in every interval inside it, from t_1 to t_2, shall differ from nat^{n-1} less than by a proposed quantity. Therefore nat^{n-1} is the velocity up to the instant t. It may be shewn in the same way that it is the velocity on from that instant. Hence the motion is elementally uniform and nat^{n-1} is the velocity at the instant t.

COMPOSITION OF VELOCITIES.

A velocity, as a directed quantity, or vector, is represented by a step; i.e., a straight line of proper length and direction drawn anywhere. The *resultant* of any two directed quantities of the same kind may be defined as *the resultant of the two steps which represent them.* This definition is purely geometrical, and it does not of course follow that the physical combination of the two quantities will produce this geometrical resultant. In the case of velocities, however, we may now prove the following important proposition.

When two motions are compounded together, the velocity in the resultant motion is at every instant the resultant of the velocities in the component motions.

Let oA, oB be velocities in the component motions at a given instant, oC their resultant. Let also oa, ob be mean velocities of the component motions during a certain interval; then we know that their resultant oc is the mean velocity of the resultant motion during that interval, because the mean velocity is simply the step taken in the interval divided by the length of the interval, and the step taken in the resultant motion is of course the resultant of the steps taken in the component motions.

Now because oA and oB are velocities in the component motions at a certain instant, we know that an interval can be found, ending at that instant, so that the mean velocities oa and ob, for every interval inside it, differ from oA, oB respectively less than a proposed quantity; so, therefore, that Aa and Bb are always both less than the proposed quantity. Now Cc is the resultant of Aa and Bb, and the greatest possible length of Cc is the sum of the lengths of Aa and Bb. We can secure, therefore, that Cc shall be less than a proposed quantity, by making Aa and Bb each less than *half* that quantity.

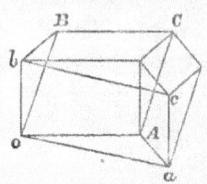

We can therefore find an interval ending at the given instant, every mean velocity inside which differs from oC less than by a proposed quantity, however small. Consequently oC is the velocity of the resultant motion up to the given instant.

It may be shewn in the same way that if oA and oB are velocities in the component motions on from the given instant, then oC is the velocity in the resultant motion on from the given instant; and therefore that when they are velocities *at* the instant, oC is the velocity at the instant in the resultant motion.

It is easy to shew in a similar manner that *when a moving point has a velocity in any position, its parallel projection has also a velocity in that position, which is the projection of the velocity of the moving point.* For let OV be the velocity of the moving point at a certain instant, OM its mean velocity in a certain interval, and let ov, om be their projections. Then the greatest possible ratio of vm to VM is that of the major axis of an ellipse,

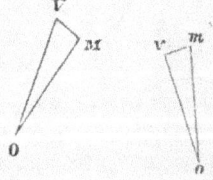

which is the parallel projection of a circle in the plane OVM, to the diameter of that circle. In order therefore, to make vm less than a proposed length, we have only to make VM less than a length which is to the proposed one as the diameter of the circle to the major axis of its projection. Hence an interval can be taken such that the mean velocity of the projected motion for every included interval shall differ from ov less than by a proposed quantity; or ov is the instantaneous velocity of the projected motion.

For an example of the last proposition, we may consider the simple harmonic motion, which is an orthogonal projection of uniform circular motion on a line in its plane. The velocity of p is na where a is the radius of the circle, and it is in the direction tp. The horizontal component of this is the velocity of m. The horizontal component is

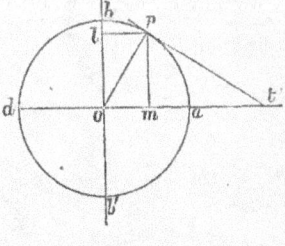

$$- na \cos ptm = - na \sin aop$$

$$= - na \sin (nt + \epsilon)$$

$$= na \cos (nt + \epsilon + \tfrac{1}{2}\pi).$$

Hence if $\qquad s = a \cos (nt + \epsilon),$

we find $\qquad \dot{s} = na \cos (nt + \epsilon + \tfrac{1}{2}\pi),$

and the rule is the same as in circular motion. The velocity is evidently $= n \cdot pm$, by which representation the changes in its magnitude are rendered clear. The same result may be obtained by means of the tangent to the harmonic curve, p. 43.

The velocity in elliptic harmonic motion may be found either by composition of two simple harmonic motions, or directly by projection from the circle. We thus find that when

$$\rho = \alpha \cos (nt + \epsilon) + \beta \sin (nt + \epsilon),$$

then $\quad \dot{\rho} = n\alpha \cos (nt + \epsilon + \tfrac{1}{2}\pi) + n\beta \sin (nt + \epsilon + \tfrac{1}{2}\pi),$

or the rule is the same as in the last case or in uniform circular motion. The result may also be stated thus: the velocity at the point p is n times oq the semiconjugate diameter.

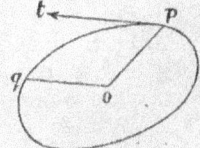

In the parabolic motion $\rho = \alpha + t\beta + t^2\gamma$ we may now see that
$$\dot{\rho} = \beta + 2t\gamma.$$

And in general if
$$\rho = \alpha + t\beta + t^2\gamma + t^3\delta + \ldots + t^n v,$$

then we shall find
$$\dot{\rho} = \quad \beta + 2t\gamma + 3t^2\delta + \ldots + nt^{n-1} v,$$

the rule being to *multiply each term by the index of t and then reduce this index by unity.* Thus we can always find the velocity when the position-vector is a rational integral function of t.

FLUXIONS.

A quantity which changes continuously in value is called a *fluent.* It may be a numerical ratio, or *scalar* quantity (capable of measurement on a scale); or it may be a *directed* quantity or vector; or it may be something still more complex which we have yet to study. In the first case the quantity, being necessarily continuous because it changes continuously, can only be adequately specified by a length drawn to scale, or by an angle; and we may always suppose an angle to be specified by the length of an arc on a standard circle. Let one end of the length which measures the quantity be kept fixed, then as the quantity changes the other end must move. *The velocity of that end is the rate of change of the quantity.* Thus we may say that water is poured into a reservoir at the rate of x gallons per minute. Let the contents of the reservoir be represented on a straight line, so that every centimeter stands for a gallon; and let the change in these contents be indicated by moving one end of the line. Then this end will move at the rate of x centimeters per minute. If w is the number of gallons in the reservoir, it is also the distance of the moveable end of the line from the fixed end, and the velocity of this moveable end is therefore $\dot{w}$. Thus we have $\dot{w} = \dot{x}$.

This rate of change of a fluent quantity is called its *fluxion,* or sometimes, more shortly, its *flux.* It appears from the above considerations that a flux is always to be conceived as a *velocity;* because a quantity must be continuous to be fluent, must therefore be specified either by a line or an angle (which may be placed at the centre of a standard circle and measured by its arc) and rate of change of a length measured on a straight line or circle means velocity of one end of it (if the other be still) or difference of velocity of the two ends.

The flux of any quantity is denoted by putting a dot over the letter which represents it.

If a variable angle *aop* be placed at the centre of a circle of radius unity, and the leg *oa* be kept still; the velocity of *p* will be the flux of the circular measure of the angle (since $ap : oa = $ circular measure, and $oa = 1$). This is called the *angular velocity* of the line *op.* When the angular velocity is uniform, it is the circular measure of the angle described in one second.

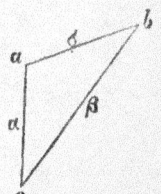

When one end of a vector is kept still, the flux of the vector is the velocity of the other end. Thus if ρ represent the vector from the fixed point *o* to the moving point *p,* $\dot{\rho}$ is the velocity of *p.* But when both ends move, the flux of the vector is the difference of their velocities. Thus if

$$\sigma = ab = ob - oa = \beta - \alpha,$$

then $$\dot{\sigma} = \dot{\beta} - \dot{\alpha}.$$

The rate of change of the vector *ab* is the velocity of *b* compounded with the reversed velocity of *a.*

DERIVED FUNCTIONS.

When two quantities are so related, that for every value of one there is a value or values of the other, so that one cannot change without the other changing, each is said to be a *function* of the other. Thus every fluent quantity is a function of t, the number of seconds since the beginning of the time considered. For example, in parabolic motion, the position-vector $\rho = \alpha + t\beta + t^2\gamma$ is a function of the time t. Here the function is said to have an analytical expression of a certain form, which gives a rule for calculating ρ when t is known. A function may or may not have such an expression.

A varying quantity being a certain function of the time, its flux is the *derived function* of the time. Thus if $\rho = \alpha + t\beta + t^2\gamma$, we know that $\dot\rho = \beta + 2t\gamma$. Then $\beta + 2t\gamma$ is the derived function of $\alpha + t\beta + t^2\gamma$. When a function is rational and integral, we know that the derived function is got by multiplying each term by the index of t, and then diminishing that index by 1. We proceed to find similar rules in certain other cases.

The flux of a sum or difference of two or more quantities is the sum or difference of the fluxes of the quantities. This is merely the rule for composition of velocities.

Flux of a product of two quantities. Let p, q be the quantities, and let p_1, q_1 and p_2, q_2 be their values at the times t_1 and t_2 respectively. Then we have to form the quotient $p_1 q_1 - p_2 q_2 : t_1 - t_2$, cast out common factors from numerator and denominator, and then omit the suffixes. Now

$$\frac{p_1 q_1 - p_2 q_2}{t_1 - t_2} = \frac{p_1 q_1 - p_1 q_2}{t_1 - t_2} + \frac{p_1 q_2 - p_2 q_2}{t_1 - t_2} = p_1 \frac{q_1 - q_2}{t_1 - t_2} + \frac{p_1 - p_2}{t_1 - t_2} q_2 ;$$

but when we cast out common factors and omit the suffixes from the latter expression, it becomes $p\dot q + \dot p q$. Thus the flux of a product is got by *multiplying each factor by the flux of the other, and adding the results.*

This is equally true when both the factors are scalar quantities, and when one is a scalar and the other a vector. We cannot at present suppose both factors to be vector quantities, because we have as yet given no meaning to such a product.

When both factors are scalar, this result may be written in a different form. Let $u = pq$, then $\dot{u} = p\dot{q} + \dot{p}q$. Divide by u, then

$$\frac{\dot{u}}{u} = \frac{\dot{p}}{p} + \frac{\dot{q}}{q}.$$

Let now $v = pqr = ur$, then we find

$$\frac{\dot{v}}{v} = \frac{\dot{u}}{u} + \frac{\dot{r}}{r} = \frac{\dot{p}}{p} + \frac{\dot{q}}{q} + \frac{\dot{r}}{r};$$

therefore
$$\dot{v} = \dot{p}qr + p\dot{q}r + pq\dot{r};$$

and it is clear that this theorem may be extended to any number of factors.

Flux of a quotient of two quantities. Let $p : q$ be the quotient; then we have

$$\left(\frac{p_1}{q_1} - \frac{p_2}{q_2}\right) \div (t_1 - t_2) = \frac{p_1 q_2 - p_2 q_1}{q_1 q_2 (t_1 - t_2)} = \frac{p_1 q_2 - p_2 q_2}{q_1 q_2 (t_1 - t_2)} - \frac{p_2 q_1 - p_2 q_2}{q_1 q_2 (t_1 - t_2)}$$

$$= \frac{1}{q_1 q_2} \left\{ q_2 \frac{p_1 - p_2}{t_1 - t_2} - p_2 \frac{q_1 - q_2}{t_1 - t_2} \right\},$$

and the latter expression, when we cast out common factors and omit the suffixes, becomes $\dot{p}q - p\dot{q} : q^2$. If we write $u = p : q$, then $\dot{u} = \dot{p}q - p\dot{q} : q^2$, or dividing by u, that is multiplying by q and dividing by p, we find

$$\frac{\dot{u}}{u} = \frac{\dot{p}}{p} - \frac{\dot{q}}{q},$$

from which a formula for the quotient of one product by another may easily be found.

We might of course use any other letter instead of t to represent the time; and when an analytical expression is

given us, involving two or more letters, we may find its *derived function*
in regard to any one of them. Thus
of the quantity $u = x^2 + 5y^2 + 3xy$,
if x represents the time, the derived
function is $2x + 3y$; but if y represents the time, the derived function
is $10y + 3x$. If we suppose x and y to be horizontal and
vertical components of a vector $op = xi + yj$, then for every
point p in the plane there will be a value of x, a value of
y, and consequently a value of u, $= x^2 + 5y^2 + 3xy$. If we
make the point p move horizontally with velocity 1 centimeter per second, then x will represent the time, and y
will not alter; so that $\dot{u}$ will be $2x + 3y$. This is called
the flux of u with regard to x, or *the x-flux of u*; and it is
denoted by $\partial_x u$. Similarly if we make p move vertically
with the unit velocity, x will be constant, and y will
represent the time, so that $\dot{u}$ will be $3x + 10y$; this is
called the flux of u with regard to y, or the y-flux of u,
and is denoted by $\partial_y u$. The characteristic ∂ may be supposed to stand for *derived function*.

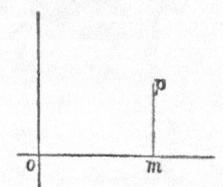

We may now prove a very general rule for finding
fluxes, namely one which enables us to find *the flux of a
function of functions*. Let x and y be two variable quantities, and let it be required to find the flux of u which is
a function of x and y; this is denoted thus: $u = f(x, y)$.
The method is the same as that used for a product. We
find

$$\frac{u_1 - u_2}{t_1 - t_2} = \frac{f(x_1, y_1) - f(x_2, y_2)}{t_1 - t_2}$$

$$= \frac{f(x_1, y_1) - f(x_2, y_1)}{t_1 - t_2} + \frac{f(x_2, y_1) - f(x_2, y_2)}{t_1 - t_2}$$

$$= \frac{x_1 - x_2}{t_1 - t_2} \cdot \frac{f(x_1, y_1) - f(x_2, y_1)}{x_1 - x_2} + \frac{y_1 - y_2}{t_1 - t_2} \cdot \frac{f(x_2, y_1) - f(x_2, y_2)}{y_1 - y_2},$$

and when we strike out common factors and omit the
suffixes in this last expression, it becomes $\dot{x}\partial_x f + \dot{y}\partial_y f$;

where f has been shortly written instead of $f(x, y)$. Or, substituting u for f, we have the formula

$$\dot{u} = \dot{x}\partial_x u + \dot{y}\partial_y u.$$

HODOGRAPH. ACCELERATION.

If a straight line ov be drawn through a fixed point o, to represent in magnitude and direction at every instant the velocity of a moving point p, the point v will describe some curve in a certain manner. This curve, so described, is called the *hodograph* of the motion of p. (ὁδὸν γράφει, it describes the way.)

Thus in the parabolic motion $\rho = \alpha + t\beta + t^2\gamma$, we have $ov = \dot{\rho} = \beta + 2t\gamma$. Hence we see that the point v moves uniformly in a straight line. *The hodograph of the parabolic motion, then, is a straight line described uniformly.* Let ab be the initial velocity; draw through b a line parallel to the axis of the parabola. Then to find the velocity at any point p, we have only to draw av parallel to the tangent at p; the line av represents the 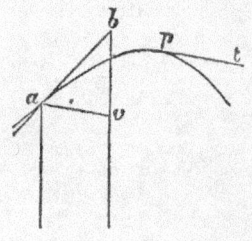 velocity in magnitude and direction. The straight line bv, described with uniform velocity 2γ, is the hodograph.

In the elliptic harmonic motion

$$\rho = \alpha \cos (nt + \epsilon) + \beta \sin (nt + \epsilon)$$

we have

$$ov = \dot{\rho} = n\alpha \cos (nt + \epsilon + \tfrac{1}{2}\pi) + n\beta \sin (nt + \epsilon + \tfrac{1}{2}\pi).$$

Thus the point v moves harmonically in an ellipse similar and similarly situated to the original path, of n times its linear dimensions, being one quarter phase in advance. As a particular case, the hodo- 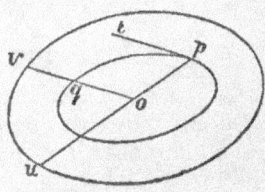 graph of uniform circular motion is again uniform circular motion. We have seen that ov is $n . oq$, where oq is the semi-conjugate diameter of op.

Of course the hodograph of every rectilinear motion is also a rectilinear motion; but in general a different one.

The velocity in the hodograph is called the *acceleration* of the moving point; thus the velocity of v is the acceleration of p. It is got from $\dot{\rho}$ in precisely the same way as $\dot{\rho}$ is got from ρ, and accordingly it is denoted by $\ddot{\rho}$. The acceleration is the flux of the velocity.

In the parabolic motion, since $\dot{\rho} = \beta + 2t\gamma$, we have $\ddot{\rho} = 2\gamma$, or *the acceleration is constant*. In the case of a body falling freely in vacuo, this constant acceleration amounts at Paris to 981 centimeters a second per second; it is called the acceleration of gravity, and is usually denoted by the letter g. It varies from one place to another, for a reason which will be subsequently explained.

In elliptic harmonic motion $\ddot{\rho}$ is to be got from $\dot{\rho}$ by the rule: Multiply by n, and increase the argument by $\frac{1}{2}\pi$. Hence

$$\ddot{\rho} = n^2 \alpha \cos (nt + \epsilon + \pi) + n^2\beta \sin (nt + \epsilon + \pi)$$
$$= - n^2 \alpha \cos (nt + \epsilon) - n^2 \beta \sin (nt + \epsilon) = - n^2\rho.$$

Thus the acceleration at p is n^2 times po; that is, it is always directed towards the centre o, and proportional to the distance from it. It is clear from the figure that the tangent at v is parallel to po; and since the velocity of v is n times ou, which is itself n times po, this velocity is n^2 times po.

Those motions in which the acceleration is constantly directed to a fixed point are of the greatest importance in physics: and we shall subsequently have to study them in considerable detail.

Acceleration is a quantity of the dimensions $[L] : [T]^2$.

THE INVERSE METHOD.

So far we have considered the problem of finding the velocity when the position is given at every instant. We shall now shew how to find the position when the velocity is given. The problem is of two kinds: we may suppose the shape of the path given, and also the *magnitude* of the velocity at every instant; or we may suppose the hodograph given. For the present we shall restrict ourselves to the first case.

Velocity being a continuous quantity, it can only be accurately given at every instant by means of a curve. Let a point *t* move along *oX* with unit velocity, and at every moment suppose a perpendicular *tv* to be set up which represents to a given scale the velocity of the moving point at that moment. Then the point *v* will trace out a curve which is called the *curve of velocities* of the moving point.

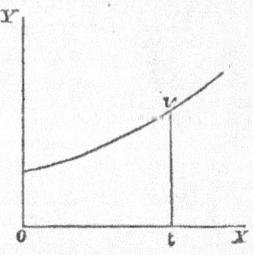

In uniform motion the curve of velocities is a horizontal straight line, *uv*. In this case we can very easily find the distance traversed in a given interval *mn*; for we have merely to multiply the velocity by the time. Now the velocity being *mu*, and the time *mn*, the distance traversed must be represented on the same scale by the area of the rectangle *umnv*.

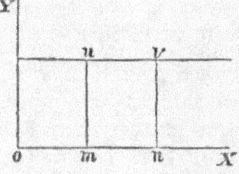

The meaning of the words *on the same scale* is this. Time is represented on *oX* on the scale of one centimeter to a second; suppose that velocity is represented on *oY* on such a scale that a centimeter in length means a velocity of one centimeter per second; then length will be represented by the area *umnv* on the scale of one square centimeter to one linear centimeter. To find the length represented by a given area, we must convert it into a rectangle standing upon one centimeter; the height of this rectangle is the length represented. The breadth, one centimeter, which thus determines the scale of representation, is called the *area-base*.

It is true also when the velocity is variable that *the distance traversed is represented by the area of the curve of velocities* (Newton). We shall prove this first for an interval in which the velocity is continually increasing; it will be seen that the proof holds equally well in the case of an interval in which it continually decreases, and as the whole time must be made up of intervals of increase and decrease, the theorem will be proved in general.

Let *uv* be the curve of velocities during an interval *mn*. Take a number of points *a, b, c* ... between *m* and *n*, and draw vertical lines *aA, bB,* *cC,* ... through them to meet the curve of velocities in *A, B, C*... The length *mn* is thus divided into a certain number of parts, corresponding to divisions in the interval of time which it represents. Through *A, B, C*... draw horizontal lines as in the figure, *ug,* *fAk, hBl,* etc. These will form as it were two staircases, one inside the curve of velocities, *ugAkBlC*..., and the other outside it, *ufAhB*... Let the horizontal line through *u* meet *nv* in *r*.

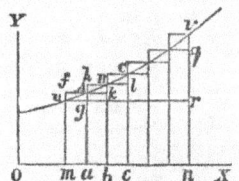

We shall now make two false suppositions about the motion of the point, one of which makes the distance traversed too small, and the other too great. *First,* suppose the velocity in the intervals *ma, ab, bc,* ... to be all through each interval what it actually is at the beginning of the interval; as the velocity is really always increasing, this supposition will make it too small, and therefore the distance traversed less than the real one. In the interval *ma,* according to this supposition, the velocity will be *mu,* and the distance traversed will be represented by the rectangle *muga.* So in the interval *ab,* the distance traversed will be *aAkb.* Hence the distance traversed in the whole interval *mn* will be represented by the area of the *inside staircase mugAkBlC...qn. Secondly,* suppose the velocity in each interval to be what it actually is at the end of the interval: then in the interval *ma* the velocity will be *aA,* and the distance traversed *mfAa.* So the distance traversed in the interval *mn* will be represented by the area of the *outside staircase mufAhB...vn.* But this supposition makes the velocity too great, excepting at the instants a, b, c ...; therefore the actual distance traversed is less than the area of the outer staircase.

It appears therefore that the distance traversed in the actual motion is represented by an area which lies between the area of the outer and the area of the inner staircase. But the area of the curve of velocities lies between these two. Therefore the difference between the area of the curve of velocities and the area which represents the distance traversed is less than the difference between the areas of the outer and inner staircases. Now this last difference is less than a rectangle, whose height is rv and whose breadth is the greatest of the lengths ma, ab, ...; for it is made up of all the small rectangles on the curve like $ufAg$. But we may divide the interval mn into as many pieces as we like, and consequently we may make the largest of them as small as we like.

It follows that there is *no* difference between the area of the curve of velocities and that which represents the distance traversed. For if there is any, let it be called δ. Divide mn into so many parts that a rectangle of the height nv, standing on any of them, shall be less than this area δ. Then we know that the difference in question is less than a rectangle of height rv standing on the greater of these parts, that is, less than δ; which is contrary to the supposition.

This demonstration indicates a method of finding the area of a curve, and, at the same time, of finding the distance traversed by a point moving with given velocity. The method is the same in the two problems (which, as we have just seen, are really the same) but has to be described in somewhat different language. For the area of the curve $umnv$, the rule is: Divide mn into a certain number of parts, and on each of these erect a rectangle whose height is the height of the curve at some point vertically over that part; then the sum of the areas of these rectangles will differ from the area of the curve by a

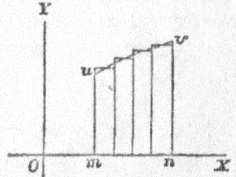

quantity which can be made as small as we like by increasing the number of parts and diminishing the largest of them. For the distance traversed during a certain interval, the rule is: Divide the interval into a certain number of parts, and suppose a body to move uniformly during each of those parts with a velocity which the actual body has at some instant during that part of the interval; then the distance traversed by the supposed body will differ from that traversed by the actual body by a quantity which can be made as small as we like by increasing the number of parts and diminishing the largest of them.

For example (Wallis), suppose the velocity at time t to be t^k, and that we have to find the space described in the interval from $t = a$ to $t = b$. Let this interval be divided into n parts in geometric progression, as follows. Let $\sigma^n = b : a$, so that $b = \sigma^n a$. Then the parts shall be the intervals between the instants $a, \sigma a, \sigma^2 a, \ldots \sigma^{n-1} a, b$. The velocities of the moving body at the beginnings of these intervals are $a^k, \sigma^k a^k, \sigma^{2k} a^k, \ldots \sigma^{(n-1)k} a^k, b^k \ldots$ Hence if a body move uniformly through each interval with the velocity which the actual body has at the beginning of that interval, it will describe the space

$$a^k(\sigma a - a) + a^k \sigma^k (\sigma^2 a - \sigma a) + \ldots + \sigma^{(n-1)k} a^k (\sigma^n a - \sigma^{n-1} a)$$

$$= a^{k+1}(\sigma - 1)(1 + \sigma^{k+1} + \sigma^{2(k+1)} + \ldots + \sigma^{(n-1)(k+1)})$$

$$= a^{k+1}(\sigma - 1) \frac{\sigma^{n(k+1)} - 1}{\sigma^{k+1} - 1} = \frac{b^{k+1} - a^{k+1}}{1 + \sigma + \sigma^2 + \ldots + \sigma^k}.$$

Now the larger n is taken, the more nearly σ approaches to unity, and consequently, the more nearly the denominator of this fraction approaches to the value $k + 1$. Thus the space described from $t = a$ to $t = b$ is $(b^{k+1} - a^{k+1}) : k+1$. By making $a = 0$ and $b = t$ in this formula, we find that the space traversed between 0 and t is $t^{k+1} : k + 1$. This agrees with our previous investigation; for if $(k+1) s = t^{k+1}$, we know that $\dot{s} = t^k$. As in the converse investigation, p. 55, it is easy to extend the method to the case in which k is a commensurable fraction; for the quotient $\sigma^{k+1} - 1 : \sigma - 1$ approaches also in that case the value $k + 1$ when σ approaches unity.

As an example, we may find the area of a parabola. Here pn^2 varies as an, or $pn = \mu \cdot an^{\frac{1}{2}}$. Thus we must put $k = \frac{1}{2}$. Then area

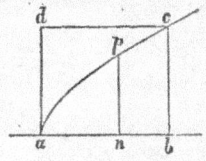

$abc = \frac{2}{3} \mu \cdot ab^{\frac{3}{2}}$, but $\mu \cdot ab^{\frac{1}{2}} = bc$,

therefore area $abc = \frac{2}{3} ab \cdot bc = $ two-thirds of the circumscribing rectangle $abcd$.

A small interval of time being denoted by δt, the approximate value of s is the sum of a series of terms like $\dot{s}\delta t$, which we may write $\Sigma \dot{s}\delta t$. The value to which this sum approaches when the number of intervals δt is increased and their size diminished, is written $\int \dot{s} dt$. Thus the equation $s = \int \dot{s} dt$ is shorthand for this statement: s is the value to which the sum of the terms $\dot{s}\delta t$ approaches as near as we like when the number of the δt is increased and their size diminished sufficiently. When the whole interval considered lies between $t = a$ and $t = b$, we indicate these *limits* of the interval thus: $s = \int_a^b \dot{s} dt$. This expression is called the *integral* of $\dot{s}$ *between the limits a and b*, or *from a to b*. Observe that although the sign $\int$ takes the place of Σ, and $\dot{s} dt$ of $\dot{s}\delta t$, yet $\int$ does not mean *sum*, nor $\dot{s} dt$ a small rectangle of breadth dt and height s. The whole expression must be taken as one symbol for a certain quantity, which indicates in a convenient way how that quantity may be calculated. $\int \dot{s} dt$ is the value to which the sum $\Sigma \dot{s}\delta t$ approaches; it is not itself a sum, but an *integral*, that is to say, a quantity which may be approximately calculated as the sum of a number of small parts.

The result obtained on the previous page may now be written thus:

$$\int_0^t t^k dt = t^{k+1} : k + 1.$$

CURVATURE.

A plane curve may be described by a point and a straight line which move together so that the point always moves along the line and the line always turns round the point. (Plücker.) Let s be the arc of the curve, measured from a fixed point a up to the moving point p, and let ϕ be the angle which the moving line (the tangent)

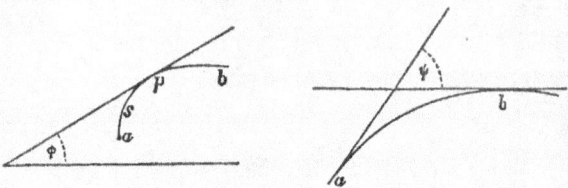

makes with a fixed line. Then the linear velocity of the point along the line is $\dot{s}$, and the angular velocity of the line round the point is $\dot{\phi}$. The ratio $\dot{\phi} : \dot{s}$ is called the *curvature* of the curve at the point p. This ratio is the s-flux of ϕ; for we know that, since ϕ is a function of s which is a function of t, $\dot{\phi} = \dot{s} \cdot \partial_s \phi$, see p. 66. Thus we may define the curvature as *the rate of turning round per unit of length of the curve.*

We may also define it independently of the idea of velocity, thus. The angle ψ between the direction of the tangents at a and b is called the *total curvature* of the arc ab; the total curvature divided by the length of the arc is called the *mean curvature* of the arc; and the curvature at any point is the value to which the mean curvature approaches as nearly as we like when the two ends of the arc are made to approach sufficiently near to that point.

In a circle of radius a, the arc $s = a\phi$; consequently $\dot{s} = a\dot{\phi}$, and $\dot{\phi} : \dot{s} = 1 : a$, or the curvature is the reciprocal of the radius. (Observe that curvature is a quantity of the dimensions $[L]^{-1}$.) It is in fact obvious that the arc of a small circle is more curved than that of a large one.

When the point stops and reverses its motion, while the line goes on, we have a *cusp* in the curve; at such a point $\dot{s} = 0$, while $\dot{\phi}$ is finite, and the curvature is infinite. When the line stops and reverses its motion, while the point goes on, we have a *point of inflexion;* at such a point $\dot\phi = 0$ while $\dot s$ is finite, and the curvature is *zero*. When both motions are reversed, we have a *rhamphoid*

cusp or *node-cusp;* the curvature is in general finite and the same on both branches.

A circle touching a curve and having the same curvature on the same side at the point of contact is called *the circle of curvature* at that point. Its radius is called the *radius of curvature*. Its centre is called the *centre of curvature*. In general the curvature is greater than that of the circle  on one side of the point, and less on the other; so that the curve crosses the circle, passing outside where its curvature is decreasing and inside where it is increasing.

If then we draw a circle to touch a curve at a point p and cut it at a point q, and then alter the radius of the circle, by moving the centre o along the normal at p, until q moves up to p, we shall obtain the circle of curvature. Hence also this circle may be described as one which has *three* points of intersection combined into one point; for the contact at p already combined two points.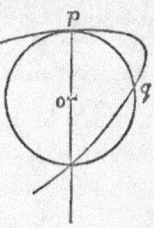

At a point of maximum or minimum curvature (like the ends of the axes of an ellipse) the curve lies wholly inside or wholly outside the circle, as in a case of ordinary contact; in such a case *four* points of intersection are combined into one.

We know that (ρ being the position-vector op) $\dot\rho$ is the velocity, and is therefore parallel to the tangent at p; and

$\dot{s}$ is the magnitude of the velocity. The quotient $\dot{\rho} : \dot{s}$, therefore, or $\partial_s \rho$, is a vector of *unit* length, parallel to the tangent at p. It is convenient to denote the s-flux of anything by a dash, just as the t-flux is denoted by a dot, so that 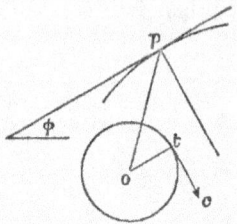 $\rho' = \partial_s \rho$. Since ot is always of unit length, t describes a circle, and the velocity of t or $\dot{\rho}'$ is the rate at which ot turns round; that is, in magnitude it is equal to $\dot{\phi}$, but its direction is tc, parallel to the inner normal at p. Now the s-flux of ρ' (which we shall write ρ'') is equal to the t-flux of ρ' divided by $\dot{s}$. But the t-flux of ρ', as we have seen, is in magnitude equal to $\dot{\phi}$. Hence ρ'', the second s-flux of ρ, is a line parallel to the inner normal at p, of length equal to the curvature at p.

When a curve does not lie in one plane (in which case it is called a *tortuous* curve), a more complex machinery is required to describe it. We must then take a point, a straight line through the point, and a plane through the straight line; and let them all move together so that the point moves along the line, the line turns *round* the point *in* the plane, and the plane turns round the line. The point is then a point on the curve, the line is the tangent at that point, and the plane is called the *osculating plane* at the point. The *curvature* is, as before, the rate of turning round of the tangent per unit of length; and, in addition, *the rate per unit of length at which the osculating plane turns round the tangent line* is called the *tortuosity*.

In this case, however, we require somewhat closer attention to determine what we mean by the rate of turning round of the tangent line. Let ot be a line of unit length always parallel to the tangent; then the point t will always lie upon a *sphere* of unit radius; but the curve not being now in one plane, t will not describe a great circle of the sphere (or circle whose plane goes through the centre). As p moves along the curve, t will describe some curve on the sphere, and the velocity of t will still be, in magnitude, the rate of turning round of ot, that is, of the tangent.

But besides this, if tc be the tangent to the path of t, the plane otc will be the plane *in which ot is turning*, that is, it will be parallel to the osculating plane. Hence tc is parallel to *the normal in the osculating plane* at p; this is called the principal normal. Since the curvature is a bending in the osculating plane, towards this normal, we may say that tc is the *direction of the curvature*.

Now in this case, just as with a plane curve, ρ' is the unit vector ot parallel to the tangent, and ρ'' is a vector parallel to tc and in length equal to the curvature. Thus ρ'' represents the curvature in magnitude and direction.

TANGENTIAL AND NORMAL ACCELERATION.

We have remarked that the t-flux of ρ is equal to the s-flux multiplied by the velocity, $\dot{s}$ or v. We may now find an expression for the second t-flux of ρ, or the acceleration, by regarding it as the flux of this product, $v\rho'$. Namely we have

$$\dot{\rho} = v\rho'$$
$$\therefore \ddot{\rho} = \dot{v}\rho' + v\dot{\rho}'.$$

But $\quad\quad\quad \dot{\rho}' = \dot{s}\rho'' = v\rho''$ (as before remarked, p. 76),

therefore $\quad\quad \ddot{\rho} = \dot{v}\rho' + v^2\rho'',$

that is to say, the acceleration $\ddot{\rho}$ may be resolved into two parts, one of which $\dot{v}\rho'$ is parallel to the tangent, and its magnitude is the rate of change in the *magnitude* of the velocity; the other $v^2\rho''$ is parallel to the (principal) normal, and its magnitude is the squared velocity multiplied by the curvature. It appears also that when the path of the moving point is tortuous, the acceleration is wholly in the osculating plane.

We may at once verify this proposition in the case of uniform motion in a circle, in which the hodograph is another circle (radius v) described uniformly. Since the two circles are described in the same time, the velocities in them must be proportional to their radii; hence the

acceleration of p, = velocity of u, : $v = v : a$, or acceleration $= v^2 : a = v^2 \times$ curvature. Thus the normal acceleration
is the same as that of a point moving with the same
velocity in the circle of curvature.

The proposition may be further illustrated by means
of the hodograph. Let ou represent the velocity of p,
and uc be the velocity of u. This
may be resolved into um in the direction of ou, which is the rate of
change in its magnitude, or $\dot{v}$; and
mc perpendicular to ou, which is ou
multiplied by its angular velocity, or
$v\dot{\phi}$, since this perpendicular velocity
may be regarded as belonging to
motion in a circle of radius v. Now since the curvature
is $\dot{\phi} : v$, it follows that $v\dot{\phi} = v^2 \times$ curvature.

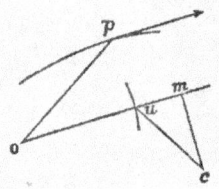

This theorem is of great use in determining the curvature of various curves.

LOGARITHMIC MOTION.

A point is said to have *logarithmic motion* on a straight
line when its distance from a fixed point on the line is
equally multiplied in equal times.

*When a quantity is equally multiplied in equal times,
its flux is proportional to the quantity itself.* Let mq, nr
be two values of such a quantity, at the times represented
by m, n; and let $mq = s$, $nr = s_1$.
Then if we move mn to the right,
keeping it always of the same
length, the ratio of s to s_1 will
remain constant; for the different intervals represented by
mn will be equal, and the
quantity is equally multiplied in equal times. We shall
have therefore $s = ks_1$, where k is this constant multiplier.
Therefore $\dot{s} = k\dot{s}_1$, and consequently $\dot{s} : \dot{s}_1 = s : s_1$. Hence
we may write $\dot{s} = ps$, where p is a constant.

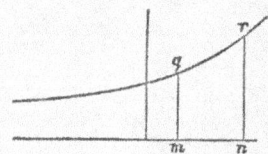

Conversely, *when the flux of a quantity is proportional
to the quantity itself, it is equally multiplied in equal times.*
For let s, s_1 be two values of the quantity, at times

separated by a given constant interval. Then we know that $\dot{s} : \dot{s}_1 = s : s_1$, or $\dot{s}s_1 - s\dot{s}_1 = 0$; that is (p. 65), the flux of the quotient $s : s_1$ is zero. Now a quantity whose flux is zero does not alter, but remains constant. Therefore $s = ks_1$, where k is constant; so that in any interval equal to the given one the quantity is multiplied by the same number k.

A quantity whose flux is always p times the quantity itself is said to increase at the logarithmic rate p.

If two quantities increase at the same logarithmic rate, their sum and difference increase at the same logarithmic rate. For if $\dot{u} = pu$, $\dot{v} = pv$, then $\dot{u} \pm \dot{v} = p(u \pm v)$.

If a quantity increases at a finite logarithmic rate, it is either never zero or always zero. For let such a quantity be zero at a and have a finite value bq at b. At the middle point c of ab it must have a value which is the geometric mean of zero and bq; that is, zero. Similarly it must be zero at the middle point of bc; and by proceeding in this way we may shew

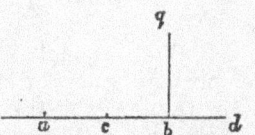

that it is zero at a point indefinitely near to any point on the left of b. If we make $bd = cb$, the value at d is a third proportional to zero and bq; that is, it is infinite. In the same way we may shew that the quantity is infinite at a point indefinitely near to any point on the right of b. It appears therefore that the quantity suddenly jumps from zero to bq and then to infinity; so that at bq the rate of increase is infinite. Hence its ratio to bq is infinite, or the logarithmic rate is infinite.

This case corresponds to the case in uniform motion when the velocity is infinite and the point is at a certain finite position at a given instant. At all previous instants it was at an infinite distance behind this position; at all subsequent instants it is at an infinite distance in front of it.

If two quantities increase at the same (finite) *logarithmic rate, they are either never equal or always equal.* For their difference is either never zero or always zero.

Let P be the result of making unity increase at the logarithmic rate p for one second; then the result of making it increase at that log. rate for t seconds is P^t when t is a whole number, for the quantity is multiplied by P in each second. It is also one value of P^t when t is a commensurable fraction, say $m : n$. For let x be its value after t seconds, then the value after nt seconds is x^n, for the quantity is multiplied by x every t seconds. But $nt = m$, and we know the result of growing for m seconds is P^m. Therefore $x^n = P^m$, or x is an nth root of P^m; that is, it is a value of P^t.

If we spread out the growth in one second over p seconds, the number expressing any velocity must be divided by p; hence if $\dot{s}$ was ps before, it must now $= s$. Hence *the result of making unity increase at the log. rate p for one second is the same as the result of making it increase at the log. rate 1 for p seconds.* Let e be the result of making unity increase at the log. rate 1 for one second; then P is a value of e^p whenever p is commensurable.

We now make this definition: the result of making unity grow at the log. rate p for t seconds is denoted by e^{pt}, and called the *exponential* of pt. The exponential coincides with one value of e to the power pt when pt is commensurable. Thus $a^{\frac{1}{2}}$ has two values, $+\sqrt{a}$ and $-\sqrt{a}$; but $e^{\frac{1}{2}}$ has only one value, the positive square root of the positive quantity e, whatever that is.

If $s = e^{pt}$, then pt is called the *logarithm* of s. The name *logarithmic rate* is given to p because it is the rate of increase of the logarithm of s.

We have an example of a quantity which is equally multiplied in equal times in the quantity of light which gets through glass. If $\frac{3}{4}$ of the incident light gets through the first inch, $\frac{3}{4}$ of that $\frac{3}{4}$ will get through the second inch, and so on. Thus the light will be multiplied by $\frac{3}{4}$ for every inch it gets through; and, since it moves with uniform velocity, it is equally multiplied in equal times.

The density of the air as we come down a hill is an example of a quantity which increases at a rate proportional to itself, for the increase of density per foot of descent is due to the weight of that foot-thick layer of air, which is itself proportional to the density,

ON SERIES.

We know that when x is less than 1, the series

$$1 + x + x^2 + \dots$$

is of such a nature that the sum of the first n terms can be made as near as we like to $\dfrac{1}{1-x}$ by taking n large enough. For the sum of the first n terms is $\dfrac{1 - x^n}{1 - x}$, and since x is less than 1, x^n can be made as small as we like by taking n large enough. The value to which the sum of the first n terms of a series can be made to approach as near as we like by making n large enough is called the *sum of the series*. It should be observed that the word *sum* is here used in a new sense, and we must not assume without proof that what is true of the old sense is true of the new one: e.g. that the sum is independent of the order of the terms. When a series has a sum it is said to be convergent. When the sum of n terms can be made to exceed any proposed quantity in absolute value by taking n large enough, the series is called divergent.

A series whose terms are all positive is convergent if there is a positive quantity which the sum of the first n terms never surpasses, however large n may be. For consider two quantities, one which the sum surpasses, and one which it does not. All quantities between these two must fall into two groups, those which the sum surpasses when n is taken large enough, and those which it does not. These groups must be separated from one another by a single quantity which is the least of those which the sum does not surpass; for there can be no quantities *between* the two groups. This single quantity has the property that the sum of the first n terms can be brought as near to it as we please, for it can be made to surpass every less quantity.

The same thing holds when all the terms are negative, if there is a negative quantity which the sum of the first n terms never surpasses in absolute magnitude.

When the terms are all of the same sign, the sum of the series is independent of the order of the terms. For let P_n be the sum of the first n terms and P the sum of the series, when the terms are arranged in one order; and let Q_n be the sum of the first n terms and Q the sum of the series, when the terms are arranged in another order. Then P_n cannot exceed Q, nor can Q_n exceed P; and P_n, Q_n can be brought as near as we like to P, Q by taking n large enough. Hence P cannot exceed Q, nor can Q exceed P; that is, $P = Q$.

When the terms are of different signs, we may separate the series into two, one consisting of the positive terms and the other of the negative terms. If one of these is divergent and not the other, it is clear that the combined series is divergent. If both are convergent, the combined series has a sum independent of the order of the terms. For let P_m be the sum of m terms of the positive series, $- Q_n$ the sum of n terms of the negative series, $P, - Q$, the sums of the two series respectively; and suppose that in the first $m + n$ terms of the compound series there are m positive and n negative terms, so that the sum of those $m + n$ terms is $P_m - Q_n$. Then $P - P_m$, $Q - Q_n$ can be made as small as we like by taking m, n large enough; therefore $P - Q - (P_m - Q_n)$ can be made as small as we like by taking $m + n$ large enough, or $P - Q$ is the sum of the compound series. It is here assumed that by taking sufficient terms of the compound series we can get as many positive and as many negative terms as we like. If, for example, we could not get as many negative terms as we liked, there would be a finite number of negative terms mixed up with an infinite series of positive terms, and the sum would of course be independent of the order.

If, however, the positive and negative series are both divergent, while the *terms* in each of them diminish without limit as we advance in the series, it is possible to make the sum of the compound series equal to any arbitrary quantity C by taking the terms in a suitable order. Suppose C positive; take enough positive terms to bring their sum above C, then enough negative terms to bring the sum below C, then enough positive terms to bring the sum again above C, and so on. We can always perform each of these operations, because each of the series is divergent; and the sum of n terms of the compound series so formed can be made to differ from C as little as we like by taking n large enough, because the terms decrease without limit.

Putting these results together, we may say that the sum of a series is independent of the order of the terms if, and only if, the series converges when we make all the terms positive.

EXPONENTIAL SERIES.

We shall now find a series for e^x, which is the result of making unity grow at the log. rate 1 for x seconds. Suppose that

$$e^x = a + bx + cx^2 + dx^3 + \dots$$

that is, suppose it is possible to find a, b, $c \dots$ so that the series shall be convergent and have the sum e^x. We will assume also (what will have to be proved) that the flux of the sum of the series is itself the sum of a series whose terms are the fluxes of the terms of the original series. Now the flux of e^x is e^x, because it grows at the logarithmic rate 1. Hence we have

$$e^x = b + 2cx + 3dx^2 + \dots$$

and this must be the same series as before. Hence

$$b = a, \quad 2c = b, \quad 3d = c, \text{ etc.}$$

Now by putting $x = 0$ we see that $a = 1$, because e^0 is the result of making unity grow for no time. Writing then for shortness Πn instead of $1 . 2 . 3 \dots n$, we find

$$e^x = 1 + x + \frac{x^2}{2} + \frac{x^3}{6} + \frac{x^4}{\Pi 4} + \dots + \frac{x^n}{\Pi n} + \dots = f(x), \text{ say.}$$

This is called the *exponential series*. We shall now verify this result by an accurate investigation.

The exponential series is convergent for all values of x. For take n larger than x; then the series after the nth term may be written thus:

$$\frac{x^n}{\Pi n} \left(1 + \frac{x}{n+1} + \frac{x^2}{(n+1) . (n+2)} + \dots \right),$$

and each term after the first two of the quantity in the brackets is less than the corresponding term of

$$1 + \frac{x}{n+1} + \frac{x^2}{(n+1)^2} + \cdots$$

which is convergent. And since it is convergent when the terms are all positive, the sum is independent of the order of the terms.

The sum of the exponential series increases at log. rate 1. Consider four quantities, x_0, x_1, x_2, x, in ascending order of magnitude. We find for the mean flux from x_1 to x_2,

$$M = \frac{f(x_1) - f(x_2)}{x_1 - x_2} = 1 + \frac{x_1 + x_2}{2} + \frac{x_1^2 + x_1 x_2 + x_2^2}{6} + \cdots$$

$$+ \frac{x_1^{n-1} + x_1^{n-2} x_2 + \cdots + x_2^{n-1}}{\Pi n} + \cdots$$

(Observe that the order of the terms has been changed, and why this is lawful.) Each term of this series is less than the corresponding term of $f(x)$, and greater than the corresponding term of $f(x_0)$. Hence the series is convergent, and its sum M lies between $f(x_0)$ and $f(x)$. And since M is finite,

$$M(x_1 - x_2) \text{ or } f(x_1) - f(x_2),$$

can be made as small as we like by making $x_1 - x_2$ small enough. Hence also $f(x) - f(x_0)$ can be made as small as we like by making $x - x_0$ small enough. Consequently we can find an interval (from x_0 to x) such that the mean flux M of every included interval (from x_1 to x_2) differs from $f(x)$ less than by a proposed quantity, however small. Therefore $f(x)$ is the flux of $f(x)$, or the sum of the exponential series increases at log. rate 1.

It follows that $f(x) = e^x$; for both quantities increase at the log. rate 1, and they are equal when $x = 0$, therefore always equal.

It appears from the investigation above, that if $f(x)$ denote the sum of a convergent series proceeding by powers of x, and $f'(x)$ the sum of the *derived* series got by taking the flux of every term; then $f'(x)$ will be the flux of $f(x)$ whenever $f'(x) - f'(y)$ can be made as small as we like by taking $x - y$ small enough; that is, when $f'x$ varies continuously in the neighbourhood of the value x.

By putting $x = 1$, we find the value of the quantity e; it is $2\cdot718281828\ldots$

THE LOGARITHMIC SPIRAL.

We may convert a step oa into a step ob by turning it through the angle aob and altering its length in the ratio $oa : ob$. But this operation may be divided into two simpler parts. From b draw bm perpendicular to oa, then $ob = om + mb$. Now we may convert oa into om by simply increasing its length in the ratio $oa : om$. Let $om : oa = x$, so that $om = x \cdot oa$. If oa' is

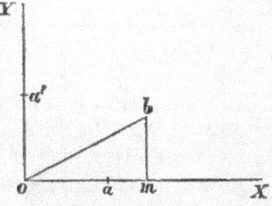

drawn perpendicular to oa, and equal to it in length, we can convert oa' into mb by multiplying it by a numerical ratio y, such that $mb = y \cdot oa'$. Now we can convert oa into oa' by turning it counter-clockwise through a right angle. Let i denote this operation; then

$$oa' = i \cdot oa.$$

Consequently $\qquad mb = y \cdot oa' = yi \cdot oa.$

And finally

$$ob = om + mb = x \cdot oa + yi \cdot oa = (x + yi)\, oa.$$

Thus the operation which converts oa into ob may be written in the form $x + yi$, where x and y are numerical ratios, and i is the operation of turning counter-clockwise through a right angle. This meaning is quite different from that which we formerly gave to the letter i. We shall never use the two meanings at the same time, in speaking of steps in one plane.

If oa be taken of the unit length, every other step ob in the plane may be represented by means of its ratio to this unit; for oa being $= 1$,

$$ob = (x + yi)\, oa = x + yi.$$

The quantities x and y will then be the components of ob parallel to oX, oY.

Since turning a step through two right angles is reversing it, $i^2 = -1$; thus i is a value of $\sqrt{(-1)}$.

The operation $x + yi$ is called a *complex number*.

The ratio $ob : oa$, which is $+\sqrt{(x^2 + y^2)}$, is called the *modulus* of the complex number $x + yi$.

If a point moves in a plane so that $\dot{\rho} = q\rho$, where q is a constant complex number, it will describe a curve which is called the *logarithmic spiral*. The velocity of the point p makes a constant angle with op and is proportional to it in magnitude. Let $q = x + yi$, then $x \cdot op$ is the component of the velocity in the direction op. If r denotes the length op, we shall have $\dot{r} = xr$, and therefore $r = ae^{xt}$, where a is the value of r at the beginning of the time. Thus the magnitude of op increases at the log. rate x. The component of velocity perpendicular to op is $yi \cdot op$; it is equal in magnitude to op multiplied by its angular velocity, or (if θ is the angle Xop) it is $op \cdot \dot{\theta}$. Hence $\dot{\theta} = y$ or the angular velocity is constant. Thus the motion of p is such that op increases at the log. rate x while it turns round with the angular velocity y. Since $\theta = yt$, while $r = ae^{xt}$, it follows that $r = ae^{k\theta}$, where $k = x : y = -\cot opt$.

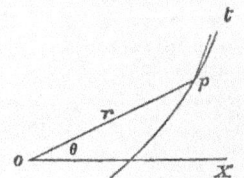

The position vector ρ of this point may be said to increase at the logarithmic rate q, because $\dot{\rho} = q\rho$. Hence we may write $\rho = ae^{qt}$, where a is the value of ρ when $t = 0$.

The meaning of e^{qt}, when q is a complex number, is *the result of making the unit step oa grow for t seconds at a rate which is got from the step at each instant by multiplying it by the complex number q.* In other words, we must make a point p start from a and move always so that its velocity is q times its position-vector; that is, its velocity must be got from the position-vector by turning it through a certain angle and altering it in a certain ratio.

We may now prove that, just as e^x is equal to the sum

of the series $f(x)$, so e^{qt} is equal to the sum of the series $f(qt)$. To make our former proof available, we have only to premise some observations on complex numbers and on series formed of them.

A complex number q alters the length of a step oa in a certain ratio (the modulus) and turns it round through a certain angle, so converting it into ob. Suppose that another complex number q_1 turns ob into oc, by altering its length in some other ratio and turning it through some other angle. Then the *product* $q_1 q$ is that complex number which turns oa into oc; it therefore multiplies oa by the product of the two ratios, and turns it through the sum of the two angles. Hence $q_1 q = q q_1$; or *the product of two complex numbers is independent of the order of their multiplication;* and *the modulus of the product is the product of the moduli.* The same thing is clearly true for any number of factors.

Instead of operating on a step with a complex number, we may operate on any plane figure whatever. The effect will be to alter the length of every line in the figure in a certain ratio, and to turn the whole figure round a certain angle. Thus the new figure will be similar to the old one. Taking for this figure a triangle, made of two steps and their sum $\alpha + \beta$, we learn that $q(\alpha + \beta) = q\alpha + q\beta$. The steps themselves may be represented by complex numbers, namely their ratios to the unit step. Hence also $(\alpha + \beta)q = \alpha q + \beta q$. Thus complex numbers are multiplied according to the same rules as ordinary numbers.

A series of complex numbers may be divided into two series by separating each term $x + yi$ into its *horizontal* (or *real*) part x and its *vertical* part yi. Neither of these parts can be greater than the modulus of the term; and therefore both parts will converge independently of the order of the terms if a series composed of the moduli converges. To change the series $f(qt)$ into the series of the moduli, we have merely to write mod. qt instead of qt; viz. the series of the moduli is $f(\text{mod. } qt)$; because the modulus of q^n is the nth power of the modulus of q.

We have before noticed that when the step ρ grows at the complex log. rate $x + yi$, its length or modulus r grows at the log. rate x. Hence ρ is either never zero or always zero.

It may now be proved successively that the series $f(qt)$ is convergent; that if t_0, t_1, t_2, t are four quantities in ascending order of magnitude, the mean flux

$$M = \frac{f(qt_1) - f(qt_2)}{t_1 - t_2}$$

differs from $qf(qt)$ by a complex number whose horizontal and vertical parts are severally less than the corresponding parts of $qf(qt) - qf(qt_0)$, whose modulus may therefore be made less than any proposed quantity by making $t - t_0$ small enough; and consequently that the flux of $f(qt)$ is $qf(qt)$. Hence it follows that $f(qt) = e^{qt}$, because they both grow at the log. rate q, and are both equal to 1 when $t = 0$.

When the velocity of p is always at right angles to op, the logarithmic spiral be-comes a circle, and the quantity q is of the form yi. Suppose the motion to commence at a, where $oa = 1$, and the logarithmic rate to be i; that is, the velocity is to be always perpendicular to the radius vector and represented by it in magnitude. Then $op = e^{it}$.

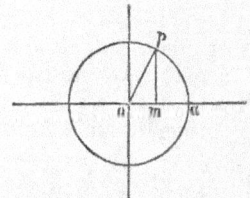

Now the velocity of p being unity in a circle of unit radius, the angular velocity of op is unity, and therefore the circular measure of aop is t. But

$$op = om + mp = \cos t + i \sin t.$$

Therefore　　　　　$e^{it} = \cos t + i \sin t,$

Euler's extremely important formula, from which we get at once the two others,

$$\cos t = \tfrac{1}{2}(e^{it} + e^{-it}), \quad i \sin t = \tfrac{1}{2}(e^{it} - e^{-it}).$$

Moreover, on substituting in these formulæ the ex-ponential series for e^{it} and e^{-it}, and remembering that $i^2 = -1$, we find series for $\cos t$ and $\sin t$, namely,

$$\cos t = 1 - \tfrac{1}{2}t^2 + \frac{t^4}{|1\underline{4}} - \frac{t^6}{|1\underline{6}} + \cdots$$

$$\sin t = t - \tfrac{1}{6}t^3 + \frac{t^5}{|1\underline{5}} - \frac{t^7}{|1\underline{7}} + \cdots$$

The formula $e^i = \cos 1 + i \sin 1$ may be graphically verified by construction of the several terms of the series

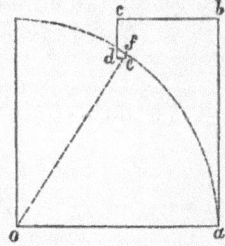

$$1 + i + \tfrac{1}{2}i^2 + \tfrac{1}{6}i^3 + \cdots$$

The first term is oa; then $ab = i$, $bc = \tfrac{1}{2}i \cdot ab$, $cd = \tfrac{1}{3}i \cdot bc$, $de = \tfrac{1}{4}i \cdot cd$, $ef = \tfrac{1}{5}i \cdot de$, and so on. The rapid convergence of the series becomes manifest, and the point f is already very close to the end of an arc of length equal to the radius.

QUASI-HARMONIC MOTION IN A HYPERBOLA.

It is sometimes convenient to use the functions $\tfrac{1}{2}(e^x - e^{-x})$, called the *hyperbolic sine* of x, hyp. sin x, or hs x, and $\tfrac{1}{2}(e^x + e^{-x})$, called the *hyperbolic cosine* of x, hyp. cos x, or hc x. They have the property $\mathrm{hc}^2 x - \mathrm{hs}^2 x = 1$. Thus whenever we find two quantities such that the difference of their squares is constant, it may be worth while to put them equal to equimultiples of the hyperbolic sine and cosine of some quantity: just as when the *sum* of their squares is constant, we may put them equal to equimultiples of the ordinary sine and cosine of some angle.

The flux of hc x is $\dot{x}$ hs x and the flux of hs x is $\dot{x}$ hc x, as may be immediately verified.

The motion $\rho = \alpha\,\mathrm{hc}\,(nt + \epsilon) + \beta\,\mathrm{hs}\,(nt + \epsilon)$ has some curious analogies to elliptic harmonic motion. Let $ca = \alpha$, $cb = \beta$, then $cm = ca \cdot \mathrm{hc}\,(nt + \epsilon)$, $mp = cb \cdot \mathrm{hs}\,(nt + \epsilon)$, so that $\dfrac{cm^2}{ca^2} - \dfrac{mp^2}{cb^2} = 1$, or $mp^2 : ma \cdot ma' = cb^2 : ca^2$. The curve having this property is called a *hyperbola*. We see at once that

$$\dot{\rho} = n\alpha \text{ hs } (nt + \epsilon) + n\beta \text{ hc } (nt + \epsilon) = n . cq, \text{ say;}$$

then
$$cp + cq = (\alpha + \beta) e^{\theta},$$

and $cp - cq = (\alpha - \beta) e^{-\theta}$ where $\theta = nt + \epsilon$. Thus pq is parallel to ab, and cn (where n is the middle point of pq) is parallel to ab'. Moreover $pn . cn = \frac{1}{4}$ product of lengths of $\alpha + \beta$ and $\alpha - \beta = \frac{1}{4} cx . cy$. Hence it appears that the

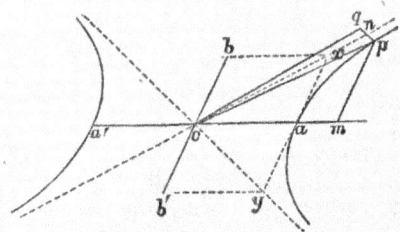

further away p goes from cy, the nearer it approaches cx, and *vice versa*. The two lines cx, cy which the curve continually approaches but never actually attains to, are called *asymptotes* (ἀσύμπτωται, not falling in with the curve). It is clear that the curve is symmetrically situated in the angle formed by the asymptotes, and therefore is symmetrical in regard to the lines bisecting the angles between them, which are called the *axes*. It consists of two equal and similar branches; though the motion here considered takes place only on one branch.

The acceleration $\ddot{\rho} = n^2 \rho$; thus it is always proportional to the distance from the centre, as in elliptic harmonic motion, but directed *away* from the centre. The lines cp, cq, are *conjugate semidiameters* of the hyperbola, as are ca, cb. Each bisects chords parallel to the other, as the equation of motion shews. The locus of q is a hyperbola having the same asymptotes, called the *conjugate* hyperbola.

The hyperbola is central projection of a circle on a horizontal plane, the centre of projection being above the lowest, but lower than the highest, point of the circle. Let b, a be highest and lowest points of the circle, v the centre of projection, am the projection of ab and pm of qn which is perpendicular to ab.   We find

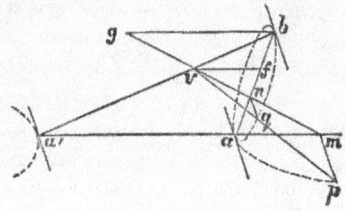

$$an \; : \; am = nf \; : \; vf \, ;$$

also $$nb \; : \; gb \; = nf \; : \; vf,$$

and $$gb \; : \; a'm = fb \; : \; af \, ;$$

multiply these three together, then

$$an \, . \, nb \; : \; am \, . \, a'm = nf^2 \, . \, fb \; : \; vf^2 \, . \, af.$$

But $$pm^2 \; : \; an \, . \, nb = pm^2 \; : \; qn^2 = af^2 \; : \; nf^2 \, ;$$

therefore $$pm^2 \; : \; am \, . \, a'm = af \, . \, fb \; : \; vf^2,$$

the property noticed above.

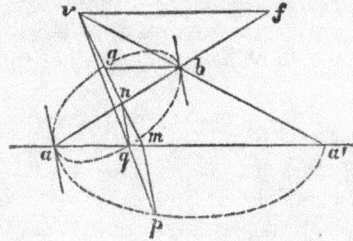

Making a change in the figure, the same process shews that the ellipse is central projection of a circle which is wholly below the centre of projection.

These three central projections of the circle, ellipse, parabola, and hyperbola, are called *conic sections;* being plane sections of the cone formed by joining all the points of a circle to a point v.

CHAPTER III. CENTRAL ORBITS.

THE THEOREM OF MOMENTS.

THE *moment* of the finite straight line *pt* about the point *o* is twice the area of the triangle *opt*. Its magnitude is the product of the length *pt* and the perpendicular on it from *o*.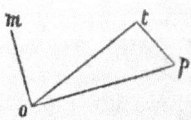

Every plane area is to be regarded as a directed quantity. It is represented by a vector drawn perpendicular to its plane, containing as many linear centimeters as there are square centimeters in the area. The vector must be drawn towards that side of the plane from which the area appears to be gone round counter clockwise. Thus *om* is the vector representing twice the area *opt*, *p* being the near end of *pt* and *m* on the upper side of the plane *opt*.

The sum of the moments of two adjacent sides of a parallelogram about any point is equal to the moment of the diagonal through their point of intersection. That is, triangle *oad* = *oac* + *oab*; each triangle being regarded as a vector, in the general case when *o* is out of the plane *abcd*. Taking first the special case of *o* in the plane, we observe that *oad* = *ocd* + *cad* + *oac*; but *ocd* + *cad* = *oab*, because the 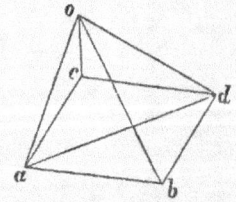 height of *oab* is the sum of the heights of *ocd* and *cad*, while all three stand on the same base *ab* or *cd*. Therefore *oad* = *oab* + *oac*.

Next, suppose o to be out of the plane. Then the vector representing oab will be a line am perpendicular to the plane oab, which may be re- 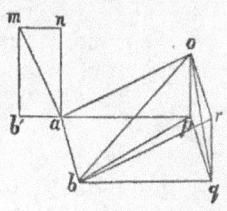 solved into components an, nm, of which an is perpendicular to the plane of the parallelogram and nm parallel to that plane. Now an represents on the same scale the projection pab of oab on the plane $abcd$, and nm its projection opq on the perpendicular plane. For the triangles oab, pab, opq, being on the same base ab or pq and having the heights respectively ao, ap, po, which are proportional to am, an, nm, must have their areas propor- tional to the lengths of these lines.

Suppose, then, the vector representing each of the areas oab, oac, oad to be resolved into components per- pendicular and parallel to the plane; the theorem will be proved if it is true separately for the components perpen- dicular and for those parallel to the plane. Now for the perpendicular components the theorem has been already proved, because they represent the triangles pab, pac, pad, which are projections on the plane $abcd$ of oab, oac, oad.

For the components parallel to the plane, observe that mn represents opq; it is at right angles to ab and proportional to 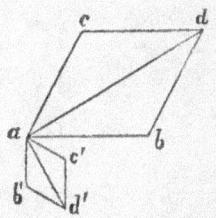 the product of ab by op the dis- tance of o from the plane. Hence the components parallel to the plane are lines ab', ac', ad' respectively at right angles to ab, ac, ad, and pro- portional to their lengths multiplied by the distance of o from the plane. Thus the figure $ab'c'd'$ is merely $abcd$ turned through a right angle and altered in scale; whence it is obvious that $ad' = ab' + ac'$.

Thus the proposition is proved in general. It is well worth noticing, however, that the proof given for the special case of o in the plane applies word for word and symbol for symbol to the general case, if only we interpret

+ and *sum* as relating to the composition of vectors. Thus *oad = ocd + cad + oac*, or *one face of a tetrahedron is equal to the vector-sum of the other three faces.* It is, of course, the sum of their projections upon it; and the components of their representative vectors which are parallel to its plane are respectively perpendicular and proportional to *oa, ad, do,* so that their vector sum is zero. Again, *ocd + cad = oab,* because the height of *oab* is the vector-sum of the heights of *ocd* and *cad.*

For proving theorems about areas, the following consideration is of great use. We have seen that *the projection of an area on any plane is represented by the projection of its representative vector on a line at right angles to the plane.* In fact, the angle *oap* between the two planes is equal to the angle *man* between the two lines respectively perpendicular to them; if we call this angle θ, the projection of the area A is $A \cos \theta$, and the corresponding projection of the line of length A is also $A \cos \theta$. Now it is easy to see that if the projections of two vectors on every line whatever are equal, then the two vectors are equal in magnitude and direction. Hence it follows that *if the projections of two areas on any plane whatever are equal, then the areas are equal in magnitude and aspect.* For example, the areas *oad* and *oac + oab* (figure on p. 92) are such that their projections on any plane are equal; this projection is, in fact, the case of the theorem of moments in which *o* is in the plane *abcd.* Hence the general theorem may be deduced in this way from that particular case.

PRODUCT OF TWO VECTORS.

On account of the importance of the theorem of moments, we shall present it under yet another aspect. The area of the parallelogram *abdc* may be supposed to be generated by the motion of *ab* over the step *ac*, or by the motion of *ac* over the step *ab*. Hence it seems natural to speak of it as the *product* of the two steps *ab, ac*. We have been accustomed to identify a rectangle with the product of its two sides, when their lengths only are taken into account; we shall now make just such an extension of the meaning of a product as we formerly made of the meaning of a sum, and still regard the parallelogram contained by two steps as their product, when their directions are taken into account. The magnitude of this product is $ab \cdot ac \sin bac$; like any other area, it is to be regarded as a directed quantity.

Suppose, however, that one of the two steps, say ac, represents an area perpendicular to it; then to multiply this by ab, we must naturally make that area take the step of translation ab. In so doing it will generate a volume, which may be regarded as the product of ac and ab. But the magnitude of this volume is ab multiplied by the area into the sine of the angle it makes with ab, that is, into the *cosine* of the angle that ac makes with ab. This kind of product therefore has the magnitude $ab \cdot ac \cos bac$; being a volume, it can only be greater or less; that is, it is a *scalar* quantity.

We are thus led to two different kinds of product of two vectors ab, ac; a *vector product*, which may be written $V \cdot ab \cdot ac$, and which is the area of the parallelogram of which they are two sides, being both regarded as steps; and a *scalar product*, which may be written $S \cdot ab \cdot ac$, and which is the volume traced out by an area represented by one, when made to take the step represented by the other.

Now the moment of ab about o is $V \cdot oa \cdot ab$; that of ac is $V \cdot oa \cdot ac$; and that of ad is $V \cdot oa \cdot ad$, which is $Voa \cdot (ab + ac)$. Hence the theorem tells us that

$$V \cdot oa \, (ab + ac) = V \cdot oa \cdot ab + V \cdot oa \cdot ac \, ;$$

or if, for shortness, we write $oa = \alpha$, $ab = \beta$, $ac = \gamma$, the theorem is that

$$V\alpha \, (\beta + \gamma) = V\alpha\beta + V\alpha\gamma.$$

We may state this in words thus: *the vector product is distributive*. And in this form the proposition may be seen at once in the figure on p. 93, if we make $ab = \alpha$, $ap = \beta$, $po = \gamma$; it asserts that

$$\text{area } abqp + \text{area } pqro = \text{area } abro,$$

and this is obviously true of their projections on any plane.

The corresponding theorem for the scalar product, that $S\alpha (\beta + \gamma) = S\alpha\beta + S\alpha\gamma$, is obvious if we regard α as an area made to take the steps β, γ.

But there is a very important difference between a vector product and a product of two scalar quantities. Namely, the *sign* of an area depends upon the way it is gone round; an area gone round counter-clockwise is positive, gone round clockwise is negative. Now if $V \cdot ab \cdot ac =$ area $abdc$, we must have by symmetry $V \cdot ac \cdot ab =$ area $acdb$, and therefore $V \cdot ac \cdot ab = -V \cdot ab \cdot ac$, or $V\gamma\beta = -V\beta\gamma$. Hence *the sign of a vector product is changed by inverting the order of the terms.* It is agreed upon that $V\alpha\beta$ shall be a vector facing to that side from which the rotation from α to β appears to be counter-clockwise.

It will be found, however, that $S\alpha\beta = S\beta\alpha$, so that the scalar product of two vectors resembles in this respect the product of scalar quantities.

MOMENT OF VELOCITY OF A MOVING POINT.

The flux of the moment of velocity of a moving point p about a fixed point o is equal to the moment of the acceleration about o. For suppose that during a certain interval of time the velocity has changed from $\dot\rho$ to $\dot\rho_1$, so that $\dot\rho_1 - \dot\rho$ is the change of the velocity; then the sum of the moments of $\dot\rho$ and $\dot\rho_1 - \dot\rho$ is equal to the moment of $\dot\rho_1$, that is the moment of the change in the velocity is equal to the change in the moment of velocity. Dividing each of these by the interval of time, we see that the moment of the mean flux of velocity is equal to the mean flux of the moment of velocity, during any interval. Consequently the moment of acceleration is equal to the flux of the moment of velocity.

The same thing may be shewn in symbols, as follows, supposing the motion to take place in one plane. We may write $\rho = re^{i\theta}$, where r is the length of op, and θ the angle Xop. Then $\dot{\rho} = \dot{r}e^{i\theta} + r\dot{\theta}.ie^{i\theta}$, or the velocity consists of two parts, $\dot{r}$ along op and $r\dot{\theta}$ perpendicular to it. The

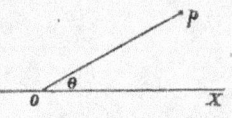

moment of the velocity is the sum of the moments of these parts; but the part along op (radial component) has no moment, and the part perpendicular (transverse component) has moment $r^2\dot{\theta}$. Next, taking the flux of $\dot{\rho}$, we find for the acceleration the value

$$\ddot{\rho} = \ddot{r}e^{i\theta} + \dot{\theta}\dot{r}.ie^{i\theta} + \dot{r}\dot{\theta}.ie^{i\theta} + r\dot{\theta}^2.i^2e^{i\theta} + r\ddot{\theta}.ie^{i\theta}$$

$$= (\ddot{r} - r\dot{\theta}^2)\,e^{i\theta} + (2\dot{r}\dot{\theta} + r\ddot{\theta})\,ie^{i\theta}.$$

Or the acceleration consists of a radial component $\ddot{r} - r\dot{\theta}^2$, and a transverse component $2\dot{r}\dot{\theta} + r\ddot{\theta}$. The moment of the acceleration is r times the transverse component, namely $2r\dot{r}\dot{\theta} + r^2\ddot{\theta}$. But this is precisely the flux of the moment of velocity $r^2\dot{\theta}$.

Observe that the radial acceleration consists of two parts, $\ddot{r}$ due to the change in magnitude of the radial velocity, and $-r\dot{\theta}^2$ due to the change in direction of the transverse velocity.

We may also make this proposition depend upon the flux of a vector product. The moment of the velocity is $V\rho\dot{\rho}$, and the moment of the acceleration is $V\rho\ddot{\rho}$; we have therefore to prove that $V\rho\ddot{\rho}$ is the rate of change of $V\rho\dot{\rho}$. Now upon referring to the investigation of the flux of a product, p. 64, the reader will see that every step of it applies with equal justice to a product of two vectors, whether the product be vector or scalar. In fact, the only property used is that the product is distributive. Hence the rate of change of $V\alpha\beta$ is $V\dot{\alpha}\beta + V\alpha\dot{\beta}$. (Observe that the order of the factors must be carefully kept.) Applying this rule to $V\rho\dot{\rho}$, we find that its rate of change is $V\dot{\rho}\dot{\rho} + V\rho\ddot{\rho}$. Now the vector product of two parallel vectors is necessarily zero, because they cannot include any area; thus $V\dot{\rho}\dot{\rho} = 0$. Therefore $\partial_t(V\rho\dot{\rho}) = V\rho\ddot{\rho}$. This demonstration does not require the motion to be in one plane.

C.

The moment of velocity about any point is equal to twice the rate of description of areas about that point. When the motion is in a circle, twice the area aop being equal to $r^2\theta$, and r constant, its flux is $r^2\dot\theta$, the moment of velocity. In any other path aq, having the same angular velocity, the area described in the same time is oaq, and the mean flux of area in the two cases is oap and oaq respectively divided by the time. The ratio of their difference to either of these is the 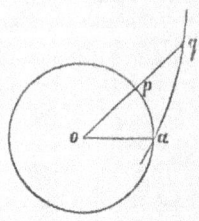 ratio of apq to oap or oaq, which is approximately the ratio of pq to op or oq, and can be made as small as we like by taking p near enough to a. Thus the mean fluxes in the two cases approach one another without limit as they approach the true fluxes; or the true fluxes are equal. Hence twice the rate of description of areas is always $r^2\dot\theta$, the moment of velocity.

When the acceleration is always directed towards a fixed point o, the moment of velocity is constant, and equal areas are swept out by the radius vector in equal times. If the acceleration of p passes through o, its moment about o is zero; consequently the flux of the moment of velocity is zero, or that moment is constant. Because it is constant in direction, the path is a plane curve; for the plane containing op and the velocity has always to be perpendicular to a fixed line. Because it is constant in magnitude, the rate of description of areas is also constant, or, which is the same thing, equal areas are swept out in equal times.

The following is Newton's proof of this proposition.

Let the time be divided into equal parts, and in the first interval let the body describe the straight line AB with uniform velocity. In the second interval, if the velocity were unchanged, it would go to c; if $Bc = AB$;

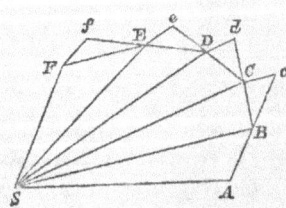

so that the equal areas ASB, BSc would be completed in equal times.

But when the body arrives at B, let a velocity in the direction BS be communicated to it. The new velocity of the body will be found by drawing cC parallel to BS to represent this addition, and joining BC. At the end of the second interval, then, the body will be at C, in the plane SAB. Join SC, then area $SCB = ScB$ (between same parallels SB and Cc) $= SBA$.

In like manner, if at C, D, E, velocities along CS, DS, ES are communicated, so that the body describes in successive intervals of time the straight lines CD, DE, EF, etc., these will all lie in the same plane ; and the triangle SCD will be equal to SBC, and SDE to SCD, and SEF to SDE.

Therefore equal areas are described in the same plane in equal intervals; and, *componendo*, the sums of any number of areas $SADS$, $SAFS$, are to each other as the times of describing them.

Let now the number of these triangles be increased, and their breadth diminished indefinitely; then their perimeter ADF will be ultimately a curved line; and the instantaneous change of velocity will become ultimately a continuous acceleration in virtue of which the body is continually deflected from the tangent to this curve; and the areas $SADS$, $SAFS$, being always proportional to the times of describing them, will be so in this case. Q.E.D.

The constant moment of velocity will be called h. It is twice the area described in one second. If p be the length of the perpendicular from the fixed point on the tangent, we shall have $h = vp = r^2\dot\theta$. A path described with acceleration constantly directed to a fixed point is called a *central orbit*, and the fixed point the *centre of acceleration*. *In a central orbit*, then, *the velocity is inversely as the central perpendicular on the tangent*, for $v = h : p$, and *the angular velocity is inversely as the squared distance from the centre*, for $\dot\theta = h : r^2$.

RELATED CURVES.

Inverse. Two points p and q so situated on the radius of a circle that $cp . cq = ca^2$, are called *inverse points* in regard to the circle.

If p moves about so as to trace out any curve, q will also move about, and trace out another curve; either of these curves is called the *inverse* of the other in regard to the circle.

The inverse of a circle is in general another circle; but it coincides with its inverse when it cuts the circle of inversion at right angles, and the inverse is a straight line when it passes through the centre of inversion. We know that

$$cp . cq = ct^2,$$

which proves the second case; the first is easily derived from it; and the third follows from the similarity of the triangles cmp, cqb, which gives

$$cp . cq = cm . cb,$$

which is constant and therefore $- cd^2 = ca^2$.

In the second case the circle clearly makes equal angles with cpq at p and q. In general, two inverse curves make equal angles with the radius vector at corresponding points. For we can always draw a circle to touch the first curve at p and to pass through q; such a circle is then its own inverse, and makes equal angles at p and q with cpq. Moreover it touches the second curve at q, for as two points of intersection coalesce at p, their two inverse points coalesce at q. Hence the two inverse curves make equal angles with cpq.

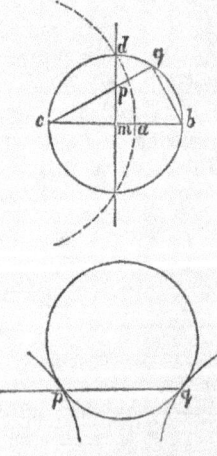

Pedal. The locus of the foot of the perpendicular from a fixed point on the tangent to a curve is called the *pedal* of the curve in regard to that point.

Let two tangents to the curve intersect in p, ct, ct' be the perpendiculars on them.

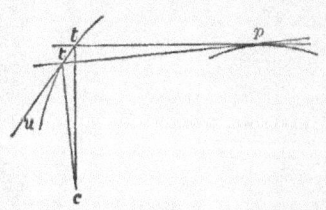

Because the angles ctp, $ct'p$ are right angles, a circle on cp as diameter will pass through tt'. Now let the two tangents coalesce into one ; then p will become a point on the curve, and $t't$ will become tangent to the pedal, and also to the circle on cp as diameter. Therefore the angle $ctu = cpt$, where tu is tangent to the pedal at t.

Reciprocal. The inverse of the pedal of a curve, in regard to the same point, is called the *reciprocal* curve.

Let s be the inverse point to t, and sn the tangent to the locus of s. We know that tu and sn make equal angles with cst ; therefore

$$csn = ctu = cpt.$$

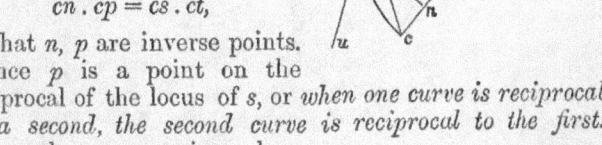

Thus the triangles csn, cpt are similar, cns is a right-angle,

and $cn : cs = ct : cp$,

or $cn . cp = cs . ct$,

so that n, p are inverse points. Hence p is a point on the reciprocal of the locus of s, or *when one curve is reciprocal to a second, the second curve is reciprocal to the first.* Hence the name, reciprocal.

We shall now shew that the reciprocal of a circle is always one of the conic sections. For this purpose it is necessary first to prove a certain property of these curves.

Two points s and h in the major axis of an ellipse, such that $sb = hb = ca$, and consequently that $cs^2 = ch^2 = ca^2 - cb^2$, are called the *foci* of the curve. Draw pm perpendicular to the axis from any point p of the curve, and take

$$cn \;:\; cm = cs \;:\; ca,$$

so that $cn . ca = cm . cs.$

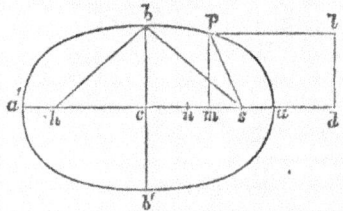

Then we shall prove that $sp = an$. For

$$sp^2 = sm^2 + pm^2 = (cs - cm)^2 + \frac{cb^2}{ca^2}(ca^2 - cm^2)$$

$$= (cs - cm)^2 + ca^2 - cs^2 - cm^2 + cn^2 = -2cs . cm + ca^2 + cn^2$$

$$= -2ca . cn + ca^2 + cn^2 = an^2.$$

Similarly $hp = na'$. Therefore $sp + hp = aa'$, or *the sum of the focal distances of any point on the ellipse is equal to the major axis.* If we take

$$cd \;:\; ca = ca \;:\; cs = cm \;:\; cn,$$

we shall have $na \;:\; md = ca \;:\; cd,$

and since $pl = md$, we have

$$sp \;:\; pl = ca \;:\; cd = cs \;:\; ca.$$

The ratio $cs \;:\; ca$ is called the *eccentricity* of the ellipse, and sometimes denoted by the letter e, so that $sp = e . pl$. The line dl is called the *directrix*.

Thus we see that the ellipse is the locus of a point whose distance from a fixed point (the focus) is in a constant ratio to its distance from a fixed line (the directrix). The distance from the focus is less than that from the directrix.

A precisely similar demonstration applies to the hyperbola; the points s and h being so taken that

$$cs = hc = ce,$$

and consequently $cs^2 = ca^2 + cb^2$. Then

$$sp^2 = sm^2 + pm^2 = (cs - cm)^2 + \frac{cb^2}{ca^2}(cm^2 - ca^2)$$

$$= (cs - cm)^2 + cn^2 - cm^2 - cs^2 + ca^2 = an$$

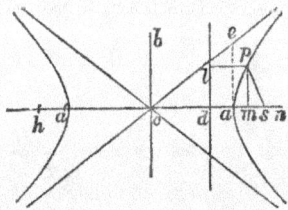

as before. So $hp = a'n$, and $hp - sp = aa'$. On the other branch we should find $sp - hp = aa'$, or *the difference of the focal distances of any point on the hyperbola is equal to the major axis.* Taking $cd : ca = ca : cs$, and drawing pl perpendicular to dl, we find as before that

$$sp : pl = cs : ca.$$

Thus in the hyperbola also the distance from the focus is in a constant ratio to the distance from the directrix dl, but the ratio in this case is greater than unity.

In the parabola we know that pm^2 varies as am; take a point s on the axis so that $pm^2 = 4as \cdot am$. Then

$$sp^2 = sm^2 + pm^2 = sm^2 + 4as \cdot am$$

$$= sm^2 + 4as \cdot sm + 4as^2 = dm^2,$$

if $da = as$. Hence $sp = pl$, or the parabola is the locus of a point whose distance from the focus s is *equal* to its distance from the directrix.

We can now prove that *the reciprocal of a circle is a conic section, of which the centre of reciprocation is a focus.*

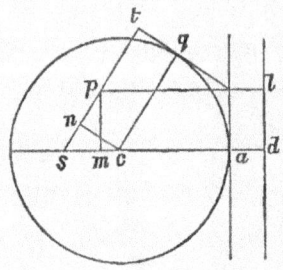

Let s be the centre of reciprocation, st perpendicular to the tangent qt of the circle. Then the reciprocal curve of the circle is inverse to the locus of t; and the size of the circle of inversion will evidently affect only the size, not the shape, of the curve. Let d be the inverse point to c, then if

$$sp \cdot st = sc \cdot sd,$$

p will be a point on the reciprocal curve. Now

$$sc \cdot sd = sp \cdot st = sp \, (sn + cq) = sm \cdot sc + sp \cdot ca$$

(since $sp : sm = sc : sn$); or $sp \cdot ca = sc \, (sd - sm) = md \cdot sc.$

Therefore $sp : pl = sc : ca$, or the locus of p is a conic section having s for focus, dl for corresponding directrix, and $sc : ca$ for eccentricity. Hence if s is within the circle this conic is an ellipse, if on the circumference a parabola, if outside the circle a hyperbola.

Since the reciprocal is the inverse of the pedal, and

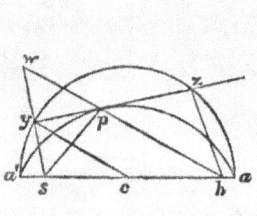
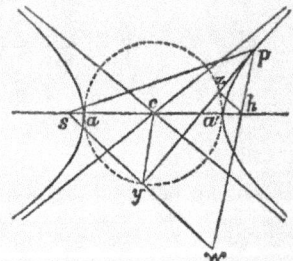

the inverse of a circle is a circle except when it passes through the centre of inversion, it follows that the pedal of a conic section in regard to a focus is a circle in the case of the ellipse and hyperbola, and a straight line in the case of the parabola. We may prove this independently thus. The tangents to an ellipse or hyperbola make equal angles with the focal distances r, $r_{,}$, for since $r \pm r_{,}$ is constant, $\dot{r} = \mp \dot{r}_{,}$; now $\dot{r}$ is the component of velocity of p along sp, and $\dot{r}_{,}$ along hp, and these being equal in magnitude, it follows that $spy = hpz$. Produce hp to w, making $pw = ps$, so that $hw = aa'$. Then sw is perpendicular to py which bisects the angle spw. Hence sy is $\frac{1}{2}sw$, and $sc = \frac{1}{2}sh$, therefore $cy = \frac{1}{2}hw = ca$, or the locus of y is the circle on aa' as diameter. This is called the auxiliary circle.

ACCELERATION INVERSELY AS SQUARE OF DISTANCE.

When the acceleration is directed to a fixed point, the hodograph is the reciprocal of the orbit turned through a right angle about the fixed point. Let py be tangent to the orbit, s the fixed point, su the velocity at p, sy perpendicular to py. Then we know that

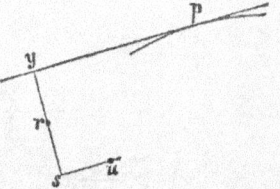

$$su . sy = h,$$

which is constant. Hence if we mark off sr on sy, so that $sr = su$, we shall have $sr . sy = h$, and therefore the locus of r is the reciprocal of the orbit. But the locus of u is the locus of r turned through a right angle.

When the acceleration is inversely as the square of the distance from the fixed point, the hodograph is a circle (Hamilton). Let the acceleration $f = \mu : r^2$, so that $fr^2 = \mu$. We know that $r^2\dot{\theta} = h$, therefore $f : \dot{\theta} = \mu : h$, or the acceleration is proportional to the angular velocity. Now the acceleration is the velocity in the hodograph, whose direction is that of the radius vector in the orbit; so that the angular velocity, which is the rate at which the radius vector turns round, is also the rate at which the tangent to the hodograph turns round. Since then the velocity in the hodograph is in a constant ratio to the rate at which its tangent turns round, the curvature of the hodograph is constant and equal to $h : \mu$. Therefore the hodograph is a circle of radius $\mu : h$.

Hence it follows directly that when the acceleration is inversely as the squared distance, the orbit is a conic section having the centre of acceleration for a focus.

Now we have

$$ys \cdot sy' = as \cdot sa' = ca^2 - cs^2 = cb^2 ;$$

and moreover $ys \cdot v = h$; whence $sy' : v = cb^2 : h$. Hence the auxiliary circle is to the hodograph (in linear dimensions) as $cb^2 : h$; or $h \cdot ca : \mu = cb^2 : h$; or $h^2 : \mu = cb^2 : ca$. If $l'sl$ be drawn through s perpendicular to the major axis, ll' is called the *latus rectum;* and we have

$$sl^2 : cb^2 = as \cdot sa' : ca^2 = cb^2 : ca^2,$$

or $sl = cb^2 : ca$. Hence $h^2 : \mu$ is the semi-latus rectum.

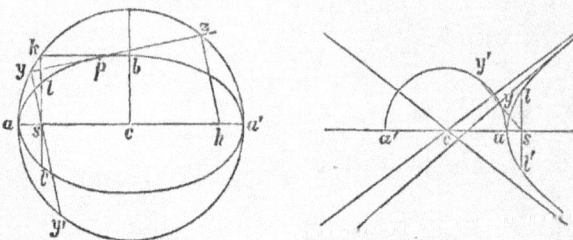

The periodic time T in the ellipse is to be found from the consideration that h is the area described in two seconds, and the area of the ellipse ($\pi \cdot ca \cdot cb$, orthogonal projection of area of circle, $\pi \cdot ca^2$) is described in T seconds.

Hence $T = 2\pi \cdot ca \cdot cb : h;$

but $cb^2 = h^2 \cdot ca : \mu$ or $cb = h \cdot \sqrt{ca} : \sqrt{\mu}.$

Therefore $T = 2\pi \cdot ca^{\frac{3}{2}} : \sqrt{\mu}$ or $\mu T^2 = 4\pi^2 \cdot ca^3.$

Consequently, in different orbits, if μ is the same, the square of the periodic time varies as the cube of the major axis.

Kepler stated three laws as the result of observation of the planets. 1st, each planet describes about the Sun areas proportional to the times. 2nd, each planet moves in an ellipse with the Sun in one focus. 3rd, the squares of the periodic times of different planets are to one another as the cubes of the major axes of their orbits. From these laws Newton deduced, 1st, that the acceleration of each planet is directed towards the Sun; 2nd, that the acceleration of each planet is inversely as the square of its distance from the Sun; and 3rd, that the acceleration of different planets is inversely as the square of their distances from the Sun, since μ is constant.

Kepler's laws and these deductions from them are however only approximately true.

ELLIPTIC MOTION.

Motion in an ellipse with acceleration always directed to one focus is called, *par excellence, elliptic motion.*

The angle *asp* is called the *true anomaly,* and is denoted by θ.

If *qpm* be drawn perpendicular to the axis, meeting the auxiliary circle in q, the point q is called the *eccentric follower* of p. Since the area *asp* is the orthogonal projection of *asq*, the latter is always proportional to it, 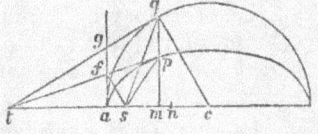 and therefore to the time; therefore q moves in the auxiliary circle[1] with acceleration always tending to s.

The angle *acq* is called the *eccentric anomaly,* and is denoted by u.

The *mean angular velocity* is also called the *mean motion,* and is denoted by n. The angle nt is called the *mean anomaly.*

It is clear that

$$nt \; : \; 2\pi = \text{area } asp \; : \; \pi . ca . cb = \text{area } asq \; : \; \pi . ca^2.$$

Now $\quad asq = acq - scq = \tfrac{1}{2} u . ca^2 - \tfrac{1}{2} cs . qm$

$$= \tfrac{1}{2} u . ca^2 - \tfrac{1}{2} e . ca . ca \sin u = \tfrac{1}{2} ca^2 (u - e \sin u).$$

---

[1] On circular orbits with acceleration to a fixed point or points, see Sylvester, Astronomical Prolusions, *Phil. Mag.* 1866.

Therefore $\qquad nt = u - e \sin u;$

an equation connecting the mean and eccentric anomalies.

The tangents at p and q meet the axis in a point t such that $cm . ct = ca^2$. Let them meet the tangent at a in f, g respectively. Then the tangents fa, fp to the ellipse subtend equal angles[1] at the focus s, and the tangents ga, gq to the circle subtend equal angles at the centre c. Consequently angle $asf = \frac{1}{2}\theta$, and $acg = \frac{1}{2}u$. We find therefore $\tan\frac{1}{2}\theta = af : as$, and $\tan\frac{1}{2}u = ag : ac$, so that

$$\tan\tfrac{1}{2}\theta : \tan\tfrac{1}{2}u = af . ac : as . ag = \sqrt{(1-e^2)} : 1-e.$$

Therefore $\qquad \tan\tfrac{1}{2}\theta = \sqrt{\dfrac{1+e}{1-e}} . \tan\tfrac{1}{2}u;$

an equation connecting the true and eccentric anomalies.

We know that $sp = an$, if $cn = e.cm$; so that, denoting sp by r, we have $r = a(1 - e \cos u)$, which gives the distance in terms of the eccentric anomaly, and a the semi-major axis.

LAMBERT'S THEOREM.

The time of getting from a point p to a point q in an elliptic orbit may be expressed in terms of the chord pq, and the sum of the focal distances $sp + sq$; a result which is called Lambert's Theorem. The following proof is due to Prof. J. C. Adams.

Let r, r' be the two focal distances, u, u' the eccentric anomalies, k the length of the chord. Regarding the chord as the projection of the corresponding chord of the auxiliary circle, we see that its horizontal component is $a(\cos u - \cos u')$ and its vertical component is

$$a\sqrt{(1-e^2)}(\sin u - \sin u');$$

for the vertical component is reduced by the projection in the ratio $a : b$, which is $1 : \sqrt{(1-e^2)}$. Hence

$$k^2 = a^2(\cos u - \cos u')^2 + a^2(1-e^2)(\sin u - \sin u')^2$$
$$= 4a^2 \sin^2\tfrac{1}{2}(u-u') \sin^2\tfrac{1}{2}(u+u')$$
$$\qquad\qquad + 4a^2(1-e^2)\sin^2\tfrac{1}{2}(u-u')\cos^2\tfrac{1}{2}(u+u')$$
$$= 4a^2 \sin^2\tfrac{1}{2}(u-u')\{1 - e^2\cos^2\tfrac{1}{2}(u+u')\}.$$

[1] Because $cm . ct = ca^2$, it is easy to shew that $ta : am = ts : an$, and therefore that $tf : fp = ts : sp$, so that sf bisects the angle asp.

Now let $u - u' = 2a$, and let β be such an angle that $e \cos \frac{1}{2}(u + u') = \cos \beta$. Then $k = 2a \sin \alpha \sin \beta$.

Moreover,

$$r + r' = 2a \left\{ 1 - \tfrac{1}{2} e \left(\cos u + \cos u' \right) \right\}$$
$$= 2a \left\{ 1 - e \cos \tfrac{1}{2}(u - u') \cos \tfrac{1}{2}(u + u') \right\}$$
$$= 2a \left(1 - \cos \alpha \cos \beta \right).$$

Therefore

$$r + r' + k = 2a \left(1 - \cos \theta \right) \text{ if } \theta = \beta + \alpha,$$

and $\quad r + r' - k = 2a \left(1 - \cos \phi \right) \text{ if } \phi = \beta - \alpha.$

Now $nt = u - u' - e(\sin u - \sin u')$
$$= 2\alpha - 2e \sin \tfrac{1}{2}(u - u') \cos \tfrac{1}{2}(u + u')$$
$$= 2\alpha - 2 \sin \alpha \cos \beta = \theta - \sin \theta - (\phi - \sin \phi).$$

Thus nt is expressed in terms of θ and ϕ, which are themselves expressed in terms of $r + r'$, k, and a. Because $nT = 2\pi$, it follows that $n^2 a^3 = \mu$; so that the time is given in terms of $r + r'$, k, a, and μ, the acceleration at unit distance.

The angle α is half the angle subtended at the centre by the corresponding chord of the auxiliary circle.

If, keeping the focus and the near vertex fixed, we make the major axis of the ellipse very large, while the points p, q remain in the neighbourhood of the focus; the ellipse will approximate to a parabolic form, and the angles u, u' will become very small; so therefore will α and β, and consequently θ and ϕ. Hence we shall have approximately

$$nt, \; = \theta - \sin \theta - (\phi - \sin \phi), \; = \tfrac{1}{6}(\theta^3 - \phi^3),$$
$$r + r' + k, \; = 2a(1 - \cos \theta), \qquad = a\theta^2,$$
$$r + r' - k, \; = 2a(1 - \cos \phi), \qquad = a\phi^2,$$

and $n^2 a^3 = \mu$ always. Therefore

$$6t \sqrt{\mu} = (r + r' + k)^{\frac{3}{2}} - (r + r' - k)^{\frac{3}{2}},$$

with an approximation which becomes closer the larger a is taken, and which becomes exact when a is infinite, or the ellipse becomes a parabola. This, therefore, is the form of Lambert's theorem for the parabola. An analogous theorem for the hyperbola will be found in the paper referred to[1].

GENERAL THEOREMS.   THE SQUARED VELOCITY.

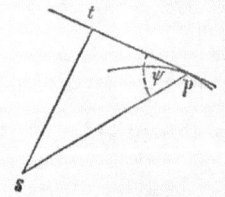

In general, if a point p be moving with acceleration f always tending from s, the resolved part of the acceleration along the tangent is $f \cos spt = f \cos \psi$, say; therefore $\dot{v} = f \cos \psi$. Now the resolved part of the velocity v along sp is $\dot{r}$, so that $\dot{r} = v \cos \psi$. It follows therefore that $f\dot{r} = v\dot{v} = \partial_t(\frac{1}{2}v^2)$. If the acceleration f depends only on the distance, so that f is a function of r, we may be able to find $\int f\dot{r}dt$ or $\int f dr$, and thence $\frac{1}{2}v^2$ to which it is equal. Suppose, for example, that $f = \mu r^{-n}$, then $(n-1)\int f dr = -\mu r^{-n+1} +$ some constant c, or $\frac{1}{2}(n-1)v^2 + \mu r^{-n+1} = c$. Since $vp = h$, this equation gives us a relation between r and p which determines the form of the orbit.

In the elliptic motion we have $\frac{1}{2}v^2 = \mu r^{-1} + c$, the acceleration being *towards* the focus, and the constant c may be determined by means of the velocity at the extremity of the minor axis, where $r = a$ and $vb = h$. Here $\frac{1}{2}h^2 = \frac{1}{2}v^2b^2 = \mu a^{-1}b^2 + cb^2$, but we know that $h^2 = \mu a^{-1}b^2$, therefore $c = -\frac{1}{2}\mu a^{-1}$ and the formula becomes

$$\tfrac{1}{2}(v^2 + \mu a^{-1}) = \mu r^{-1}.$$

The analogous formula for the hyperbola is

$$\tfrac{1}{2}(v^2 - \mu a^{-1}) = \mu r^{-1},$$

which may be found by considering the velocity at an infinite distance, when the point may be regarded as moving along the asymptote.

Since a parabola may be regarded as an infinitely long ellipse or as an infinitely long hyperbola, we find the corresponding formula for that case by making a infinite in

---

[1] *Messenger of Mathematics*, 1877.

either of the two others, viz. $\frac{1}{2} v^2 = \mu r^{-1}$; in this case the velocity at an infinite distance is zero.

We see then that when a point starts from the position p at a distance r from s, and moves with acceleration μr^{-2} always tending to s; if the velocity at starting is $\sqrt{(2\mu r^{-1})}$, the path will be a parabola; if less than this, an ellipse with semi-major axis given by the formula $\mu a^{-1} = 2\mu r^{-1} - v^2$; if greater, an hyperbola with semi-major axis given by the formula $\mu a^{-1} = v^2 - 2\mu r^{-1}$. The major axis of the orbit depends only on the velocity, not at all on the direction, of starting.

A special case of elliptic motion is that in which, the direction of starting being in the line sp, the ellipse reduces itself to a straight line. The foci then coincide with the extremities of the major axis, the eccentricity $e = 1$, and the motion is the projection on aa' of motion in the circle with acceleration tending to a. Writing x for ap, and u for the angle acq, we have

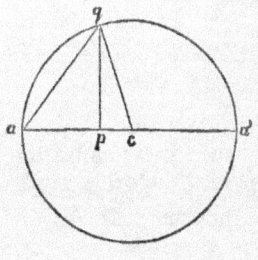

$$x = a\,(1 - \cos u), \quad nt = u - \sin u = 2\sin^{-1}\frac{x}{2a} - \frac{\sqrt{(2ax - x^2)}}{a};$$

from which equations it may be verified with a little trouble that $\ddot{x}x^2 = -n^2a^3$. It follows that if from a point p in the ellipse a point be started with the velocity belonging to the elliptic motion in the direction sp, and have always an acceleration μr^{-2}, it will ascend to a point r such that $sr = aa'$, and then return to p with the same velocity; so that *the velocity at any point of the ellipse is that due to a fall from the circle rk.* If we join ph and produce it to cut the ellipse at q, we have $ph = pr$, $hq + sq$

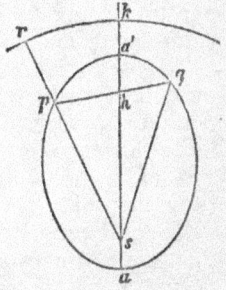

$= aa' = sr$, and therefore $pq + sq = pr + sr$. Hence if an ellipse be described with the foci s, p, to touch the circle at r, it will pass through q and touch the ellipse pq at that point (since both tangents must make equal angles with sq, hq). Thus all the orbits which can be described from p with given velocity touch an ellipse having foci s, p and major axis $sr + pr$. Or, in purely geometrical terms, given a focus, one point, and the length of the major axis of an ellipse, its envelop is the ellipse here specified.

In the case of the further branch of a hyperbola described with acceleration *from* the focus, the velocity is that due to a fall out from the circle rk, from r to p. We have again $ph = pr$, and $sq - hq = sr$, therefore

$$sq - pq = sr - rp,$$

or q is on a hyperbola with foci s, p, touching the circle at r and the orbit at q.

When the nearer branch is described with acceleration *to* the focus, the theorem becomes rather more complex. If a point be started from p in the direction sp, with the velocity belonging to the hyperbolic orbit, and acceleration *from s*, its velocity will approximate to a certain definite value more and more closely as it gets further and further away. If we now suppose a point to approach from an infinite distance on the other side of s, with a velocity more and more nearly equal to the same value the greater the distance from s, but now with acceleration *from s*, this point will come up to the position r (where $sr = aa'$), and there stop and go back. So that if now we reverse this process, start a point from

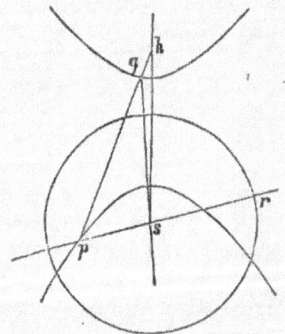

rest at r and make it fall *through infinity* to the point p, it will arrive at p with the velocity belonging to the hyperbolic orbit. We have again $ph = pr$, $sq - hq = sr$, therefore $pq + sq = pr + sr$, or q lies on an *ellipse* with foci s, p touching the circle at r and the further branch of the hyperbola at q.

Returning to the case of the ellipse, we know that if it is lengthened out until one focus goes away to an infinite distance, it will become a parabola. If however we send away the focus h, the circle rk, having a fixed centre s and a radius increasing without limit, will itself go away to infinity; and there will be no proper envelop of the different parabolic paths which pass through p. In a parabola described with acceleration towards the focus, therefore, the velocity at every point is that due to a fall from infinity; or, as we may say, the velocity in the parabola is the *velocity from infinity*.

If, holding fast the focus h, we send s away to infinity, all the lines passing through it become parallel, and their ratios unity; so that the acceleration becomes constant in magnitude and direction, and we fall back on the previously considered case of *parabolic motion*. Since ha' is then $= a'k$, the circle rk becomes the directrix; and we learn that in the parabolic motion the velocity at any point is that due to a fall from the directrix. The envelop of the orbits described by points starting from a given point with given velocity is a parabola having that point for focus and touching the common directrix at r.

GENERAL THEOREMS. THE CRITICAL ORBIT.

We have shewn that when the acceleration $f = \mu r^{-n}$, then $\frac{1}{2}(n-1)v^2 + \mu r^{1-n}$ is a certain constant, c. For convenience, suppose that $c = \frac{1}{2}(n-1)u^2$ where u is a certain velocity; then if we make r infinite, and suppose n greater than 1, r^{1-n} will be zero, and we shall have $v = u$. Hence u is the velocity at an infinite distance; and if the orbit has any infinite branch, u is the value to which the velocity of a particle going out on that branch would indefinitely approach. If however n is less than 1 or negative, r^{1-n} will be zero when r is zero, and in this case

c.   8

u is the velocity of passing through the centre of acceleration. If we draw a circle with that point as centre, and
radius a determined by the equation $\frac{1}{2}(n-1)u^2 = \mu a^{1-n}$,
then in the case $n > 1$, the velocity at every point is that
due to a fall from this circle, either directly or through
infinity; and in the case $n < 1$, the velocity is that due to
a fall from the circle either directly or through the centre :
it being understood that in passing through infinity or
through the centre the sign of μ must be changed.

Just as when $n = 2$ the parabola is a critical form of
orbit, dividing from one another the ellipses and hyperbolas, so in general, an orbit in which $u = 0$ is called a
critical orbit. When $n > 1$, the velocity at every point of
such an orbit is that due to a fall from an infinite distance
(in this case μ must be negative, or the acceleration
towards the centre); and when $n < 1$, the velocity is that
due to a fall from the centre, μ being positive and the
acceleration away from the centre. In both cases

$$\tfrac{1}{2}(n-1)v^2 = \mu r^{1-n}.$$

Since $vp = h$, we find $\frac{1}{2}(n-1)h^2 r^{n-1} = \mu p^2$; or the orbit
is of such a nature that p varies as a power of r. Conversely, in any curve in which p varies as a power of r, we
can find the acceleration with which it may be described
as a critical orbit.

Now this is the case when $r^m = a^m \cos m\theta$. For we
know that the resolved parts of the velocity of P, along r
and perpendicular to r respective-
ly, are $\dot{r}$ and $r\dot{\theta}$. Consequently
$-\tan\psi = r\dot{\theta} : \dot{r}$. But if $r^m = a^m \cos m\theta$,
we must have $r^{m-1}\dot{r} = -a^m\dot{\theta}\sin m\theta$,
so that $\cot m\theta = -r\dot{\theta} : \dot{r} = \tan\psi$.
Therefore $\cos m\theta = \sin\psi = p : r$, or
$a^m p = r^{m+1}$. Comparing this with
our previous expression, we find
$2m + 2 = n - 1$, or $m = \frac{1}{2}(n-3)$,

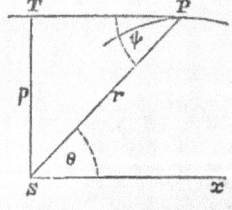

and $a^{2m} = \frac{1}{2}(n-1)h^2 : \mu$, or $a^m = h\sqrt{\dfrac{n-1}{2\mu}}$. Changing

the sign of m is equivalent to taking the inverse curve,
since it replaces r by $a^2 : r$. We subjoin a list of curves

belonging to this class, observing that each is the pedal, the inverse, and the reciprocal of another curve of the series.

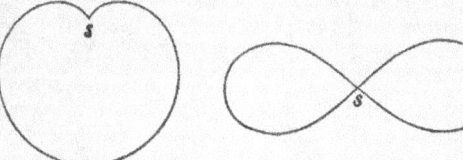

Cardioid.	Lemniscate.

$m = 1$, $n = 5$; circle passing through s.

$m = -1$, $(n = 1)$; straight line (exceptional).

$m = \frac{1}{2}$, $n = 4$; "cardioid," pedal of circle in regard to s; inverse of parabola.

$m = -\frac{1}{2}$, $n = 2$; parabola with focus at s.

$m = 2$, $n = 7$; "lemniscate," inverse and pedal of rectangular hyperbola. (hyperbola with perpendicular asymptotes).

$m = -2$, $n = -1$; rectangular hyperbola with centre at s; is its own reciprocal.

The straight line, as we know, cannot be described with acceleration to any point out of it; and in fact the case $n = 1$, which the formula points to, is an exceptional one. From $f = \mu : r$, $\partial_t (\frac{1}{2} v^2) = f\dot{r} = \mu \dot{r} : r$, we deduce[1] $\frac{1}{2} v^2 = \mu \log r$ or $h^2 = 2\mu p^2 \log r$, which is not a curve of the kind here considered.

Another exceptional case is the logarithmic spiral, in which r is proportional to p, and consequently $n = 3$, $m = 0$. A point started from a given position with the velocity from infinity and acceleration μr^{-3} will describe a logarithmic spiral, in which the only thing that can vary is the angle at which it cuts all its radii vectores. In particular, if the point start at right angles to the radius vector, it will describe a circle.

If we write $z = x + iy$, and $\zeta = \xi + i\eta$, supposing $x + iy$

---

[1] If $x = \log y$, then $y = e^x$, and $\dot{y} = \dot{x} e^x = \dot{x} y$; hence $\dot{x} = \dot{y} : y$, or $\dot{y} : y = \partial_t \log y$.

to be the position-vector of a point z in one plane, and
$\xi + i\eta$ of a point ζ in another plane, then any relation
between z and ζ will enable us to find one of these points
from the other; and if z move about describing any figure
in its plane, ζ will describe a corresponding figure in *its*
plane. Now if $\zeta = z^n$, and one of the points describe a
horizontal or vertical line, the other will describe a critical curve. For we may write $z = re^{i\theta}$, $\zeta = se^{i\phi}$, then
we shall have $se^{i\phi} = r^n e^{in\theta}$, whence $s \cos\phi = r^n \cos n\theta$, and
$s \sin\phi = r^n \sin n\theta$. Suppose then that ζ moves in a vertical line, so that ξ, $= s\cos\phi$, is kept constant, then $r^n \cos n\theta$
is constant, or z describes a critical curve. If ζ moves in
a horizontal line, so that η, $= s\sin\phi$, is kept constant, then
$r^n \sin n\theta$ is constant, which gives the same curve turned
through an angle $\pi : 2n$.

EQUATION BETWEEN u AND θ.

Let, as before, SP be denoted by r, ST by p, TP by q,
and let ϕ be the angle which PT makes with a fixed line.
The components of velocity of
T along ST and TP, are $\dot{p}$
and $p\dot{\phi}$ But the components
of velocity of T relative to P
are $q\dot{\phi}$ and $\dot{q}$ in those directions; and the velocity of P
is $\dot{s}$ along TP. Hence we
have $\dot{p} = q\dot{\phi}$, or $q = \partial_\phi p$; and
$\dot{s} - \dot{q} = p\dot{\phi}$, or

$$\partial_\phi s = p + \partial_\phi q = p + \partial_\phi^2 p.$$

This value of the radius of curvature $\partial_\phi s$ is often useful.

We may for example apply it to the hodograph of a
central orbit. Let the letters r, p, θ, ϕ, refer to the orbit,
and let r_1, p_1, θ_1, ϕ_1, mean the corresponding quantities in
regard to the hodograph. Then we know that $r_1 = v$, and
$rp_1 = r_1 p = h$, so that writing u for the reciprocal of r, or
$ur = 1$, we have $p_1 = hu$. Moreover $\phi_1 = \theta$, and the radius
of curvature of the hodograph $= f : \dot{\phi_1} = f : \dot{\theta} = f : hu^2$.

Making these substitutions, the formula becomes

$$h^2 u^2 (\partial_\theta^2 u + u) = f.$$

This formula connects the law of acceleration with the shape of the orbit, independently of the time of description.

By means of it we may prove a useful proposition relating to the effect of adding to the acceleration with which a given orbit is described a new acceleration, directed to the same centre, inversely as the cube of the distance. We shall then have $h^2 u^2 (\partial_\theta^2 u + u) = f - \mu u^3$, or if $\mu = h^2 (n^2 - 1)$, then $h^2 u^2 (\partial_\theta^2 u + n^2 u) = f$. Now let $\phi = n\theta$, then $\dot{\phi} = n\dot{\theta}$ and $\partial_\theta u = \dot{u} : \dot{\theta} = n\dot{u} : \dot{\phi} = n\partial_\phi u$, so that $\partial_\theta^2 u = n^2 \partial_\phi^2 u$. As $hu^2 = \dot{\theta}$, a change of θ into ϕ would change h into nh; now the equation is $n^2 h^2 u^2 (\partial_\phi^2 u + u) = f$. Thus in the new state of things, when the value of u is the same as before, θ is changed into $n\theta$. Therefore the same effect may be produced by letting the point move as before in its original orbit, while that orbit turns round the point s with $n - 1$ times the angular velocity of the moving point.

BOOK II. ROTATIONS.

CHAPTER I.

STEPS OF A RIGID BODY.

THERE are two kinds of motion of a rigid body which are comparatively simple, and which it is convenient to study first by themselves. The first is the motion of a body sliding about on a plane (*e.g.* a book on a table), which may be completely described by specifying the motion of a moving plane on a fixed plane. The second is the motion of a body, one point of which is fixed; which in practice is secured by a ball-and-socket joint, and which is most conveniently studied under the form of the sliding of a spherical surface on an equal spherical surface. When the centre of a sphere is very far away from the surface, the surface approximates to that of a plane. Thus the frozen surface of still water is approximately spherical, with its centre at the centre of the earth. In this way we may see that the first of our two motions is only a limiting case of the second, in which the fixed point is an infinite distance off.

As in the case of translations we shall at first attend only to the change of position or *step* which the body makes between the beginning and end of the time considered, without troubling ourselves about what has taken place in the interval.

In the case of a plane sliding on a plane, the motion is determined if we know the motion of two points a, b, or the finite line ab. So in a sphere sliding on an equal sphere, the motion is determined if we know the motion of the arc of *great circle ab*. (A *great circle* on a sphere is one whose plane passes through the centre.)

Every change of position in a plane sliding on a plane may be produced either by translation or by rotation about a fixed point. Let the straight line *ab* be moved to *a'b'*; it will be sufficient if we prove that this step can be effected in the way named, since the motion of all the rest of the plane is determined by that of *ab*. Join *aa'*, *bb'* and bisect them at right angles by the lines *co, do*. First, let these

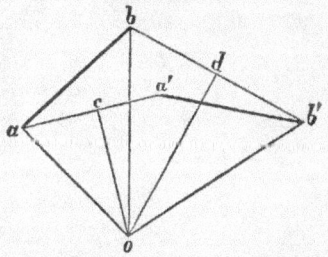

meet in *o*. Then *oa = oa'*, and *ob = ob'* ; and of course *ab = a'b'*; so that the triangles *oab, oa'b'* have their sides respectively equal, and therefore the angle *aob = a'ob'*. Hence also angle *aoa' = bob'*. Therefore if the triangle *oab* be turned round the fixed point *o*, until *oa* comes to *oa'*, *ob* by the same amount of turning will come to *ob'*, and consequently the triangle *oab* will come to coincide with *oa'b'*.

Next, suppose that the lines bisecting *aa'*, *bb'* at right angles are parallel to one another. Then *aa'*, *bb'* are parallel, and consequently either *ab* is parallel to *a'b'*, and the required step is a translation, or else they make equal angles with *aa'*, *bb'*, and one can be brought to coincide with the other by rotation round their point of intersection

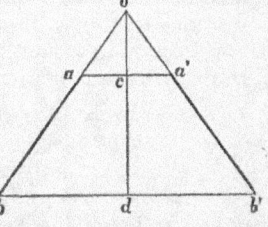

o. In the latter case the bisecting lines coincide, and the point *o* is not determined by their intersection.

Two figures which are equal and similar are called *congruent*. If they can be moved so as to coincide with each other, they are called *directly* congruent; but if one is the image of the other in a plane mirror they are said to be *inversely* congruent, or one is a *perversion* of the other. Two plane figures which are inversely congruent can be moved into coincidence by taking one of them out of its plane and turning it over; this does not make them directly congruent in regard to the plane.

It is essential to the preceding demonstration that the two triangles *oab*, *oa'b'* should be *directly* congruent. Now

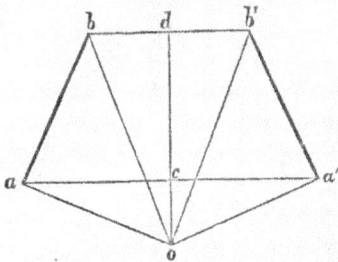

if they were inversely congruent, as in this figure, the lines bisecting *aa'* and *bb'* at right angles would coincide, contrary to the supposition.

It is to be observed that the case of translation occurs when the lines *co, do* are parallel, that is, when their point of intersection *o* has been sent off to an infinite distance. Thus a step of translation may be regarded as a step of rotation round an infinitely distant point.

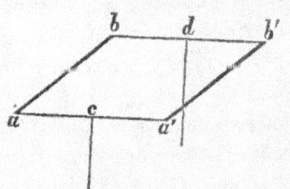

Every change of position in a sphere sliding on an equal sphere may be produced by rotation about a fixed point. The proof is exactly the same as before, except that straight lines are to be replaced by great circles of the sphere, and that the case of *co, do* being parallel does not occur; for *any* two great circles intersect in two opposite points of the sphere, say *o* and *o'*. Rotation about *o* is rotation about the axis *oo'*, therefore also about *o'*. The theorem may be also stated thus: every displacement of a body having one point fixed may be produced by rotation about an axis through that point. The fixed point is of course the centre of the sphere.

*Every displacement of a rigid body may be produced by
rotation about a fixed axis together with translation parallel
to the axis* (screw motion).   Let *a* be any point of the
body whose new position is *a'* ; then we can produce the
whole displacement by first giving the body a translation
aa', and then turning it about *a'* as a fixed point.   The
latter step can be effected by rotation about an axis
through *a'*. Now consider those points of the body which
lie in a plane perpendicular to this axis. By the rotation
they are merely turned round in that plane ; while by the
translation the plane was moved parallel to itself. Hence
the new position of this plane is parallel to its original
position. Let then the body have first a translation perpendicular to the plane, so as to bring the plane into its
new position; then the remaining displacement consists of
a sliding of this plane on itself, which may be produced
by rotation about a fixed point of it, or, which is the same
thing, about an axis perpendicular to the plane. Thus the
whole displacement is produced by rotation about that
axis, together with translation parallel to it.

If two plane polygons, which are perversions of one
another, be rolled symmetrically along a straight line, one
on each side, until the same two corresponding sides come
into contact, the result will be merely a translation of each
along the line through a distance equal to its perimeter.
Hence successive finite rotations through angles equal to
the exterior angles of a polygon about successive vertices
(taken the same way round) are equivalent to a translation
of length equal to the perimeter. By supposing one
polygon fixed, and the other to roll round it, we find that
successive rotations about the vertices through *twice* the
exterior angles will bring the plane back to its original
position.

The corresponding theorems for a spherical surface are
easily stated.

CHAPTER II.   VELOCITY-SYSTEMS.

WHEN a body is rotating about a fixed axis with angular velocity ω, every point in the body is describing a circle in a plane perpendicular to the axis, whose radius is the perpendicular distance of the point from the axis. Hence the velocity of the point is in magnitude ω times its distance from the axis, and its direction is perpendicular to the plane which contains the axis and the point.

If ab be the axis, pm perpendicular to it, the velocity of p is ω times mp perpendicular to the plane pab. If, therefore, we represent the angular velocity ω by means of a length ab marked off on the axis, the velocity of p is ab multiplied by mp, which is the moment of ab about p, being twice the area pab.

In the case of a plane figure, the rotation being about an axis perpendicular to the plane, or say about a point m in the plane (where it is cut by the axis), the velocity of any point p is $\omega \, . \, mp$ in magnitude, but perpendicular to mp; that is, it is $i\omega \, . \, mp$, the angular velocity being reckoned positive when it goes round counter clockwise.

When a body has a motion of translation, the velocity of every point in it is the same, and that is called the velocity of the rigid body. But in the case of rotation, the velocity of different points of the body is different, and we can only speak of the system of velocities, or *velocity-system*, of its different points. Still, the velocity-system due to a definite angular velocity about a definite axis is spoken of as the *rotation-velocity*, or simply the *velocity* of a rigid body which has that motion. To specify it completely we must assign its magnitude and the position of the axis; it is thus represented by a certain length marked off anywhere on a certain straight line. For it clearly does not matter on what part of the axis the length *ab* is marked off; its moment in regard to *p* will always be the velocity of *p*. A rotation-velocity, so denoted, shall be called a *spin*.

Such a quantity, which has not only magnitude and direction, but also position, is called a *rotor* (short for *rotator*) from this simplest case of it, the rotation-velocity of a rigid body. A rotor is a *localised* vector. While the length representing a vector may be moved about anywhere parallel to itself, without altering the vector, the length representing a rotor can only be slid along its axis without the rotor being altered.

Two velocity-systems are said to be *compounded* into a third, when the velocity of every point in the third system is the resultant of its velocities in the other two.

COMPOSITION OF SPINS.

The resultant of two spins *l, m* about the points *a, b* in a plane, is a spin $(l + m)$ about a point *c*, such that $l \cdot ca + m \cdot cb = 0$. For the velocities of *p* due to the two spins are $il \cdot ap$ and $im \cdot bp$, and their resultant is consequently $i(l + m)cp$; that is, it is the velocity due to a spin $l + m$ about *c*.

It should be observed that the result holds good whatever be the signs of *l, m*; but that, if their signs are different, the point *c* will be in the line *ab* produced. There is one very important exception, when the spins are equal but of opposite signs; the resultant is then a

translation-velocity.   Let the spins
be $l, -l$, then

$$il \cdot ap - il \cdot bp = il\,(ap - bp) = il \cdot ab.$$

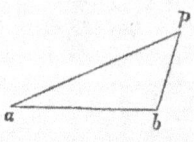

Thus the velocity of every point p is
the same, namely it is of the magnitude $l \cdot ab$ and is perpendicular to ab.

Translating these results into language relating to
axes perpendicular to the plane, we
find that the resultant of two parallel
spins l, m is a spin of magnitude
equal to their sum, about an axis
which divides any line joining them
in the inverse ratio of their magnitudes.   But the resultant of two equal
and opposite parallel spins is a trans-

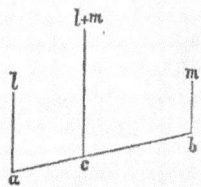

lation-velocity, perpendicular to the plane containing them,
of magnitude equal to either multiplied by the distance
between them.

It follows that if we compound a spin l with a translation-velocity v perpendicular to its axis, the effect is to
shift the axis parallel to itself through a distance $v : l$ in a
direction perpendicular to the plane containing it and the
velocity.

A translation-velocity may be regarded as a spin about
an infinitely distant axis perpendicular to it.   Hence all
theorems about the composition of translation-velocities
with spins are special cases of theorems about the composition of spins.

*The resultant of two spins about axes which meet is a
spin about the diagonal of the parallelogram whose sides
are their representative lines, of the magnitude represented by that diagonal.*   In other
words, spins whose axes meet are
compounded like vectors.   For if
ab, ac represent the two spins,
and ad is the diagonal of the
parallelogram $acdb$, the velocities
of any point p due to the two
spins are the moments of ab and
ac about p, and the resultant of

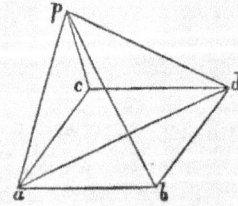

them is the moment of ad about p, that is, it is the velocity due to a spin ad.

It follows from this that the resultant of any number of spins whose axes meet in a point is also a spin whose axis passes through that point. And that if i, j, k are spins of unit angular velocity about axes oX, oY, oZ at right angles to one another, any spin about an axis through o may be represented by $xi + yj + zk$, where x, y, z are magnitudes of the component spins about the axes oX, oY, oZ.

VELOCITY-SYSTEMS. TWISTS.

If a rigid body have an angular velocity ω about a certain axis, combined with a translation-velocity v along that axis, the whole state of motion is described as a *twist-velocity* (or more shortly, a *twist*) about a certain *screw*. We may in fact imagine the motion of the body to be produced by rigidly attaching it to a nut which is moving on a material screw. The ratio $v : \omega$ is called the *pitch* of the screw; it is a linear magnitude (of dimension $[L]$ simply), and we may cut a screw of given pitch upon a cylinder of any radius. The pitch is the amount of translation which goes with rotation through an angle whose arc is equal to the radius. For our present purpose it is convenient to regard the axis of the rotation as a cylinder of very small radius, on which a screw of pitch p is cut. The screw is entirely described when its axis is given, and the length of the pitch. The angular velocity ω is called the *magnitude* of the twist.

The velocity of a point at distance k from the axis is $k\omega$ perpendicular to the plane through the axis, due to the rotation, and v parallel to the axis, due to the translation. If the resultant-velocity makes an angle θ with the axis, we shall have $\tan \theta = k\omega : v = k : p$. Thus for points very near to the axis, the velocity is nearly parallel to it; for points very far off, nearly perpendicular to it; and for points whose distance is equal to the pitch of the screw, it is inclined at an angle of 45°.

A quantity like a twist-velocity, which has magnitude, direction, position, and pitch, is called a *motor*, from the twist-velocity which is the simplest example of it, and which, as we shall see, is the most general velocity-system of a rigid body.

COMPOSITION OF TWISTS.

The resultant of any number of spins and translation-velocities is a twist. Take any point o, and let ab represent one of the spins. Then ab is equivalent to an equal spin about the parallel oc, together with a translation-velocity which is the moment of ab about o. In the same way every other spin of the system may be resolved into a spin about an axis through o and a translation-velocity. Then all the spins will have for resultant a spin about an axis through o, and all the translation-velocities will have for resultant a translation-velocity. Let os be the resultant spin, and ot the resultant translation-velocity; then ot may be resolved into om along os and mt perpendicular to it. The effect of combining the spin os with mt is to shift its axis parallel to itself perpendicular to the plane sot through a distance $mt : os$. Thus we are left with a spin about an axis parallel to os and a translation along that axis; that is to say, the resultant is a twist.

It follows, of course, that the resultant of any number of twists is also a twist. We shall now determine the axis and pitch of the resultant of two twists[1]. It is convenient to suppose in the first place that the axes of the twists intersect at right angles. Let then oX, oY be these axes, α, β the magnitudes of the twists, a, b their pitches, ϖ, p, the magnitude and pitch of the resultant twist, k the distance of its axis

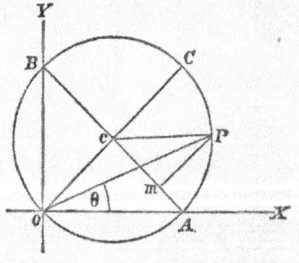

---

[1] This theory, and most of the nomenclature of the subject, are due to Dr Ball.

from the point o, θ the angle it makes with oX. Then $\alpha = \varpi \cos\theta$, $\beta = \varpi \sin\theta$, the two spins α, β about oX, oY compounding into a spin ϖ round oP. The translations due to these spins are $a\alpha$, $b\beta$, or $\varpi a \cos\theta$, $\varpi b \sin\theta$, along oX, oY. The sum of their resolved parts along OP

$$= \varpi a \cos\theta . \cos\theta + \varpi b \sin\theta . \sin\theta = \varpi \, (a \cos^2\theta + b \sin^2\theta).$$

The sum of their resolved parts perpendicular to OP

$$= \varpi a \cos\theta . \sin\theta - \varpi b \sin\theta . \cos\theta = \tfrac{1}{2}\varpi \, (a - b) \sin 2\theta.$$

The latter part shifts the axis OP parallel to itself in a direction perpendicular to the plane through a distance

$$k, = \tfrac{1}{2} \, (a - b) \sin 2\theta.$$

The former part shews that the pitch of the resultant twist

$$p, = a \cos^2\theta + b \sin^2\theta.$$

Now let a circle be drawn through o and two points A, B on oX and oY equidistant from o. The centre c is the middle point of AB. Then since θ is the angle at the circumference AoP, 2θ is the angle at the centre AcP,

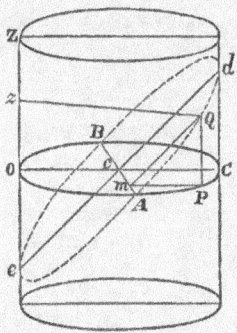

and $\sin 2\theta = Pm : cA$. If a cylinder be drawn upon this circle, a plane through AB and a point vertically over C at a distance $\tfrac{1}{2} \, (a - b)$ will cut the cylinder in an ellipse, and if Q be the point of the ellipse vertically over P we shall have $PQ = k$. For

$$PQ : Pm = Cd : Cc, \quad Pm = Cc \sin 2\theta,$$

and $Cd = \frac{1}{2}(a - b)$,

whence $PQ = \frac{1}{2}(a - b)\sin 2\theta = k.$

Hence zQ, parallel to oP, is the axis of the resultant twist.

The angle θ depends upon the magnitude of the component twists, not at all upon their pitches. By varying this angle then, we shall obtain the screws of all twists which can be got by compounding twists upon the given screws. If θ varies uniformly, the line zQ, which is parallel to OP, turns round uniformly, being always perpendicular to oZ; while the point z has a simple harmonic motion up and down oZ, whose period is equal to that of P in the circle. The surface traced out by the line zQ is called a cylindroid. It is clear that if we cut the cylindroid by a circular cylinder having oZ for axis, the section will be the bent oval previously obtained by wrapping round the cylinder two waves of a harmonic curve (p. 35). The line oZ is called the *directrix* of the cylindroid.

The pitch of each screw on the cylindroid depends only on its position and the pitches of the two component twists; to represent therefore the distribution of pitch we may attribute to these twists any absolute magnitude that we like. We shall suppose their magnitudes to be inversely proportional to the square roots of their pitches. Let oa and ob be these magnitudes, and let the pitches be represented by numbers on such a scale that the pitch of oa is $ob : oa$, then the pitch of ob is $oa : ob$, since the pitches are as $ob^2 : oa^2$. Then the translation accompanying the spin oa will be represented by $i \cdot ob$, and that accompanying ob by $i \cdot oa'$ or $i \cdot oa$ according as the two pitches are of the same or different signs. In the first case construct an ellipse, in the second a hyperbola, with oa and ob for semi-axes; then we shall shew that the translation accompanying a spin op, regarded as compounded of proper multiples of oa and ob, is $i \cdot oq$, where oq is the semi-conjugate diameter.

To prove this, we must observe that, pm and qn being drawn perpendicular to the major axis, $om : oa = nq : ob$, and $\pm on : oa = mp : ob$. For the ellipse this follows by parallel projection from the circle, in which the property is obvious; for the hyperbola we know that

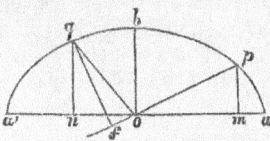

$$om = oa \cdot \text{hc}\,\phi, \quad on = oa \cdot \text{hs}\,\phi, \quad mp = ob \cdot \text{hs}\,\phi, \quad nq = ob \cdot \text{hc}\,\phi,$$

where ϕ is written for $nt + \epsilon$ of p. 89.

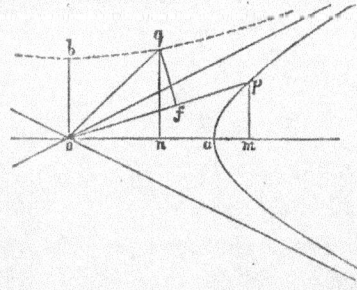

Thus the spin op being equivalent to om and mp, the translation due to om is to $i \cdot ob$ as $om : oa$, that is, it is $i \cdot nq$; and the translation due to mp is $mp : ob$ multiplied by $i \cdot oa'$ and $i \cdot oa$ in the two cases respectively, that is, it is $i \cdot on$. Hence the translation due to op is $i \cdot oq$.

If we draw qf perpendicular to op, $of : op$ will be the height k of the screw which is parallel to op, and $qf : op$ will be its pitch. Now in the harmonic or quasi-harmonic motion with acceleration towards the centre, $n \cdot po$ is the velocity at q, and fq is equal to the perpendicular from the centre on the tangent at q; therefore the rectangle $op \cdot fq$ is constant, and consequently equal to $oa \cdot ob$. Hence

$$qf : op = oa \cdot ob : op^2,$$

or *the pitch of the screw parallel to op is inversely proportional to the square of op.*

This ellipse or hyperbola is called the *pitch-conic.*

When the pitch-conic is a hyperbola, it follows that there are two screws of pitch zero, namely those which are

C. 9

parallel to the asymptotes. Thus in two cases the resultant twist is a pure spin. The distance from o of these is
1 and -1 respectively. Thus the scale on which the
pitches have been reckoned is such that the unit of length
is half the distance between the axes of pure spin. When
the pitch of the screw on oX is zero, the pitch-conic reduces to two lines parallel to oX; and there is no other
screw whose pitch is zero, except when that of oY is zero,
and then all the pitches are zero, the cylindroid reducing
to the lines through o in the plane XoY.

In order to shift the figure of the pitch-conic through
a distance k perpendicular to
its plane, we must add $ki \cdot oa$
to the translation accompany-
ing the spin oa, and $ki \cdot ob$ to
that accompanying ob. Let

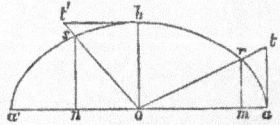

$$at = k \cdot ob, \text{ and } bt' = k \cdot oa';$$

then the new translations are ot, ot', which are still along
conjugate diameters, because by similar triangles we have

$$mr : om = k \cdot ob : oa \text{ and } on : ns = k \cdot oa' : ob;$$

whence
$$\frac{om}{oa} \cdot \frac{on}{oa'} = \frac{mr}{ob} \cdot \frac{ns}{ob},$$

which is the condition.

The resultant of two twists whose axes are anyhow
situated is a twist about some screw which belongs to a
cylindroid containing the axes of the given twists. This
cylindroid we now proceed to find, supposing the two
screws given. Find the line which meets both of their
axes at right angles; this is the
directrix of the cylindroid. Draw
a plane through one of the axes
perpendicular to the directrix, and
a line in this plane parallel to the
other axis meeting the directrix.
Let ob be the first axis, oq perpendicular to it, then $i \cdot oq$ will be
the direction of the translation
that goes with the spin about

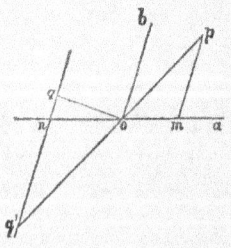

ob; let oa be parallel to the second axis, and $i\,.\,op$ the direction of the translation which together with a spin about oa is equivalent to a twist about that axis. If p be the pitch of the second screw, h the distance of its axis from o, $\tan aop = p : h$. Then the problem is to find an ellipse (or hyperbola) having oa, ob for conjugate diameters, and also op, oq. Or rather, having given that these are the directions of two pair of conjugate diameters, it is necessary to find the relative magnitudes of one pair.

For this purpose we observe that if p, q are points on the conic, $on : om = mp : nq$, or the areas onq, omp are equal. Let po meet qn in q'; then

$$\text{area } omp : \text{area } onq = om^2 : on^2$$

since they are similar. But $onq : onq' = nq : nq'$ so that $om^2 : on^2 = nq : nq'$. Given q, this determines p, so that the ratio op, oq is known. A conic described on these as semi-conjugate diameters is similar to the pitch-conic. Screws parallel to its axes compounded of the two given screws will be the oX, oY of the cylindroid.

The analytical solution is as follows[1]. Let p, q be the pitches, k_1, k_2 the distances from o the centre of the cylindroid, l, m the inclinations to oX, of the two screws, h their distance and θ the angle between them. Then from the equations

$$p = a \cos^2 l + b \sin^2 l, \qquad q = a \cos^2 m + b \sin^2 m,$$

$$k_1 = \tfrac{1}{2}(a - b) \sin 2l, \qquad k_2 = \tfrac{1}{2}(a - b) \sin 2m, $$

we have to find a, b, l, m, k_1, k_2 in terms of p, q, h, and θ. Now

$$h = k_1 - k_2 = \tfrac{1}{2}(a - b)(\sin 2l - \sin 2m)$$
$$= (a - b) \cos (l + m) \sin (l - m),$$

and

$$k_1 + k_2 = \tfrac{1}{2}(a - b)(\sin 2l + \sin 2m)$$
$$= (a - b) \sin (l + m) \cos (l - m).$$

So also

$$p - q = \tfrac{1}{2}(a - b)(\cos 2l - \cos 2m)$$
$$= (b - a) \sin (l + m) \sin (l - m),$$

[1] Ball, *Theory of Screws*, pp. 16, 17.

and $\quad p + q = a + b + \frac{1}{2}(a-b)\cos(2l + \cos 2m)$

$\qquad = a + b + (a-b)\cos(l+m)\cos(l-m)$

$\qquad = a + b + h\cos\theta \ (\text{since } l - m = \theta).$

Therefore $\qquad h^2 + (p-q)^2 = (a-b)^2\sin^2\theta,$

$\qquad (q-p)\cot\theta = k_1 + k_2,$

$\qquad p - q = h\tan(l+m) = h\tan(2l-\theta);$

whereby $a \pm b,\ k_1 \pm k_2,\ l$ and m are expressed in terms of $p,\ q,\ \lambda,$ and $\theta.$

MOMENTS.

When a straight line moves as a rigid body, the component of velocity along the line of every point on it is the same. For consider two points, a, b ; the rate of change of the distance ab is the difference of the resolved parts of the velocities of a and b along ab. If therefore the length ab does not change, this difference is zero. This component of velocity of any point on the line may be called the *lengthwise velocity* of the line.

The lengthwise velocity of a line due to a given twist is called the *moment* of the twist about 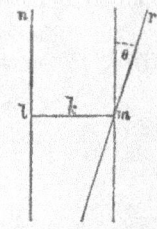 the line. Let $lm, = k,$ be the shortest distance between the axis ln of the twist and the straight line mr. It will be sufficient to determine the velocity of m along mr. Now m has the velocity $k\omega$ perpendicular to the plane mln, and $p\omega$ parallel to ln, if ω be the magnitude and p the pitch of the twist. Let θ be the angle between mr and ln, then the resolved parts of these components along mr are $-k\omega\sin\theta$ and $+p\omega\cos\theta$. Thus the moment of the twist about the line is $\omega(p\cos\theta - k\sin\theta)$.

The moment of a screw about a straight line is the moment of a unit twist on that screw about the line. Thus $p\cos\theta - k\sin\theta$ is the moment of a screw of pitch p about a line at distance k making an angle θ with its axis.

All the straight lines in regard to which a given screw has no moment, are said to form a *complex* of lines belonging to that screw. When a line belonging to the complex is moved by a twist about the screw, every point in it moves at right angles to the line.

All the lines of the complex which pass through a given point lie in a given plane, namely, the plane through the point perpendicular to its direction of motion due to a twist about the screw. This plane passes through the perpendicular from the point on the axis, and makes with the axis an angle θ, such that $\tan\theta = p : k$.

Conversely, all the lines of the complex which lie in a given plane pass through a certain point, at a distance $p\cot\theta$ from the axis along a straight line in the plane perpendicular to it. If any other line in the plane belonged to the complex, every point in the plane would move perpendicularly to the plane, and the twist would reduce to a spin about some line in the plane.

In the case when $p = 0$, or the twist reduces itself to a spin about its axis, the moment becomes $-k\sin\theta$, and can only vanish if the line meet the axis ($k = 0$), or is parallel to it ($\sin\theta = 0$), which is the same as meeting it at an infinite distance. Hence the complex reduces itself to all the lines which meet the given axis.

All the lines of the complex which meet a given straight line, not itself belonging to the complex, meet also another straight line. For, suppose the cylindroid constructed, which contains the given screw and the given straight line, considered as a screw of pitch 0. Then the pitch-conic must be a hyperbola, since there is one screw with pitch 0; this is parallel to one asymptote, and there must be another parallel to the other asymptote. Hence *every twist may be resolved into two spins, the axis of one of which is any arbitrary straight line,* not belonging to its complex. Now, since the two spins are equivalent to the twist, the lengthwise velocity of any line due to the twist is the sum of its lengthwise velocities due to the two spins; or the moment of the twist is the sum of the moments of the two spins. If then a straight line belong to the complex and meet the axis of one spin, the moments of the twist and one spin are zero, consequently the moment of the other spin is zero, or its axis meets the line. Therefore a straight line of the complex which meets the axis of one spin, meets also the axis of the other.

If however the axis of one spin belong to the complex, that of the other spin must meet it, since the moment of the twist about it is zero; but in that case it must also coincide with it, since otherwise the pitches of all screws on the cylindroid would be zero. We have then the case noticed above, in which the pitch-conic reduces to two parallel lines.

From the symmetry of the expression $- k \sin \theta$ in regard to the two straight lines concerned, we perceive that *the lengthwise velocity of a line* A *due to a unit spin about a line* B *is equal to the lengthwise velocity of* B *due to a unit spin about* A. Hence we may speak of this quantity as the *moment of the two lines*, or of either in regard to the other. We shall also define the moment of two spins as *the product of their magnitudes into the moment of their axes.* If one of the axes goes away parallel to itself to an infinite distance, and at the same time the angular velocity ω about it diminishes indefinitely, so that $k\omega = v$, the spin becomes a translation-velocity v perpendicular to that axis, making, therefore, an angle $\phi, = \frac{1}{2}\pi - \theta$, with the other axis; and the moment becomes $v\omega' \cos \phi$, if ω' is the magnitude of the finite spin. In the same way we may speak of the moment of a twist and a spin, meaning the magnitude of the spin multiplied by the moment of the twist about its axis.

Suppose the twist resolved into two spins A, B; then its moment in regard to the spin C will be the sum of the moments of the component spins. Let us combine with C a spin D, making a second twist; then the sum of the moments of the twist $A + B$ in regard to C and D will be equal to $(AC) + (BC) + (AD) + (CD)$, (where (AC) means the moment of A in regard to C), that is, it will be the sum of the moments of the twist $C + D$ in regard to A

and B. Therefore *it is independent of the way in which the second twist is resolved into two spins.*

Consider then two twists α, β, whose pitches are p, q. The moment of the first in regard to the *rotation* of the second is $\alpha\beta\,(p\cos\theta - k\sin\theta)$, and in regard to the *translation* it is $\alpha\,.\,\beta q\cos\theta$. Thus the whole moment is

$$\alpha\beta\,[(p+q)\cos\theta - k\sin\theta].$$

The quantity $(p+q)\cos\theta - k\sin\theta$ is called the moment of the two screws, or of either in regard to the other. It may be thus defined:—Let a unit twist about one screw be resolved into two spins, and let the magnitude of each of these be multiplied by the lengthwise velocity of its axis due to a unit twist about the other screw. The sum of the products is the moment of the two screws.

Hence, by making the two twists coincide, we find that the moment of a twist in regard to itself is the square of its magnitude, multiplied by twice its pitch. Since then the moment of a spin in regard to itself is zero, the moment of a twist $A+B$ is twice the moment of the spins A, B; and this is therefore the same, whatever two spins the twist is resolved into.

Now the moment of two spins in regard to one another is six times the volume of the tetrahedron which has the lines representing the spins for opposite edges. Let ab, cd be the representative lines; since each may be slid along its axis without altering the spin, let them be so placed that the shortest distance fg bisects them both. Draw through f, $a'b'$ equal and parallel to ab, bisected by f; and through g, $c'd'$ equal and parallel to cd, bisected by g. Then $a'd\,b'c$, 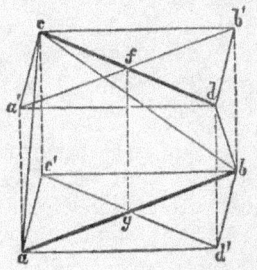 $ad'\,bc'$ are equal parallelograms, and the volume of the parallelepiped, of which they are opposite faces, is

$$\tfrac{1}{4}fg\,.\,ab\,.\,cd\sin\theta,$$

which is half the moment of ab in regard to cd. Now this parallelepiped is made up of the tetrahedra

$$abcd, \ abdd', \ abcc', \ cdaa', \ cdbb',$$

of which the last four are equal, and each of them (being one-third of height × base) is one-sixth of the parallelepiped. It follows that $abcd$ is one-third of it.

We learn then that a twist may be resolved in an infinite number of ways into two spins, but that the tetrahedron, whose opposite edges are their representative lines, is always of the same volume, namely, one-sixth of the squared magnitude of the twist multiplied by its pitch.

INSTANTANEOUS MOTION OF A RIGID BODY.

We shall prove presently that when a plane is in motion, sliding on another plane, the system of velocities at any instant is that of a spin about a certain point in the plane, called the *instantaneous centre*. As the motion goes on, the instantaneous centre in general changes continuously, describing a curve in the fixed plane and a curve in the moving plane. These curves are called *centrodes* (κέντρου ὁδός, path of the centre), and the motion is such that the centrode in the moving plane (the moving centrode) *rolls* upon the centrode in the fixed plane (the fixed centrode). Thus every motion of a plane sliding on a plane may be produced by the rolling of one curve on another; the point of contact being the instantaneous centre.

Similar theorems hold good when a body moves about a fixed point, or, which is the same thing, when a spherical surface slides upon an equal sphere. In this case the velocity-system at any instant is that of a spin about a certain line through the fixed point, called the *instantaneous axis;* or, in describing the sliding of a sphere, we may say that at any instant it is rotating about a point on the spherical surface, called the *instantaneous centre*. As the motion goes on, the instantaneous axis moves, always

passing through the fixed point, so as to describe a certain
cone, called the *fixed axode*. At the same time it traces
out *in the moving body* another cone, called the *moving
axode*. We may describe the same thing in other words
by saying that the instantaneous centre on the spherical
surface describes a fixed and a moving centrode on the
fixed and moving spheres respectively. The motion is
such that the moving cone *rolls* upon the fixed cone, and
therefore the moving spherical curve rolls upon the fixed
curve.

The most general motion of a rigid body is that of a
twist about a certain screw, called the *instantaneous screw*.
The axis of this screw, in moving about, generates two
surfaces, one fixed in space, and one moving with the body.
These surfaces are called *axodes;* being generated by the
motion of a straight line, they belong to the class of *ruled
surfaces* or *scrolls*. The motion is such that one axode
rolls and slides on the other, the line of contact being the
axis of the instantaneous screw.

Returning to the motion of a plane on a plane, we may
approximately represent it during a certain interval by considering a series of successive positions at
certain instants during the interval. We
know that the body may be moved from one
of these to the next by turning it round
a certain point.   Let a, B, C, D, E ... be
the points round which the body must be
turned in order to take it from the first
position to the second, the second to the

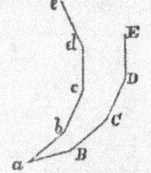

third, etc., and let b, c, d, e ... be the points in the moving
plane which successively come to coincide with B, C, D, E...
Then we can move the body through this series of positions
by rolling the polygon $abcde$ on the polygon $aBCDE$, it
being obvious that corresponding sides of them are equal.
By taking the successive positions sufficiently near to one
another, we can make this approximation as close as we
like to the actual motion of the plane ; and the nearer the
successive positions are taken, the more closely do the
polygons approximate to continuous curves which roll
upon one another.

Precisely similar reasoning may be used in the case of a sliding sphere, and of the general motion of a rigid body.

There are some difficulties in this proof, which the following exact investigation may clear up. The question is, what velocity-systems are consistent with rigidity? We shall secure that the body does not change in size or shape, if we make sure that no straight line in the body is altered in length. Let a and b be two points in the body, then the motion of b relative to a must be at right angles to ab; for its component along ab is the flux of the length ab, which has to be zero. We shall find it convenient to denote the velocity of the point a by $\dot{a}$. This being so, it is necessary and sufficient for rigidity that $\dot{b} - \dot{a}$ should be either zero, or perpendicular to ab, where a, b are *any* two points in the moving body. It follows at once that *if two velocity-systems are consistent with rigidity, their resultant is consistent with rigidity.*

Now suppose a plane to be sliding on a plane, and combine with its velocity-system a translation equal and opposite to the velocity of any point a. Then the new motion is consistent with rigidity, and the point a is at rest. Consequently the new motion is a spin about the point a. The original motion, therefore, is the resultant of this spin and of a translation equal to the velocity of a; it is therefore a spin of the same magnitude ω, about a point o situate on a line through a perpendicular to its direction of motion, at a distance such that $\dot{a} = i\omega \cdot oa$.

To determine the motion of the instantaneous centre, we must find the acceleration of any point in the plane. The instantaneous centre shall be called c in the fixed plane, and c_1 in the moving plane; and at a certain instant of time it shall be supposed to be at a point o in the moving plane. Then at that instant c, c_1, o are the same point; but $\dot{c}$ means the velocity of the instantaneous centre in the fixed plane, $\dot{c}_1$ its velocity in the moving plane, and $\dot{o}$ the velocity of o in the moving plane, which we know to be zero.

Now if p be any point in the moving plane, we know that at every instant $\dot{p} = i\omega \cdot cp$. To find the acceleration of p we must remember that the flux of cp is $\dot{p} - \dot{c}$. Therefore

$$\ddot{p} = i\dot{\omega} \cdot cp + i\omega (\dot{p} - \dot{c}) = (i\dot{\omega} - \omega^2) cp - i\omega \cdot \dot{c}.$$

Now let p coincide with o, that is (for the instant) with c. Then $\ddot{o} = -i\omega \cdot \dot{c}$, or the *acceleration of o is at right angles to the velocity of c, and equal to the product of it by the angular velocity.*

If we suppose the moving plane to be fixed, and the fixed plane to slide upon it so that the relative motion is the same, then if p_1 is the point of the fixed plane which at a given instant coincides with p in the moving plane, the velocity and acceleration of p_1 on one supposition are equal and opposite to the velocity and acceleration of p on the other supposition; also ω becomes $-\omega$. Hence we shall have $\ddot{o}_1 = + i\omega \cdot \dot{c}_1$, but $\ddot{o}_1 = -\ddot{o}$. Therefore $\dot{c}_1 = \dot{c}$, or *the velocity of the instantaneous centre in the moving plane is the same in magnitude and direction as its velocity in the fixed plane.*

Because these velocities are the same in direction, the two centrodes touch one another; and because they are the same in magnitude, the moving centrode rolls on the fixed one without sliding. For let s, s_1 be the arcs ac, bc 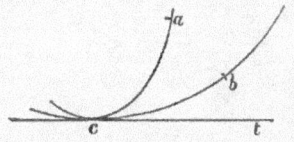 measured from points a, b which have been in contact; then $s = \dot{s}_1$, and therefore (since they vanish together) $s = s_1$.

The angular velocity ω is equal to $\dot{s}$ multiplied by the difference of the curvatures of the two centrodes. For suppose them to roll simultaneously on the tangent ct; then their angular velocities $\dot{\phi}$ and $\dot{\psi}$ will be respectively equal to their curvatures multiplied by $\dot{s}$, and the relative angular velocity will be the difference of these. When the curvatures are in opposite directions one of them must be considered negative. The same result may be obtained by calculating the flux of the acceleration of o.

Thus if r, r_1 are radii of curvature of the fixed and rolling centrodes, we have

$$\frac{\omega}{s} = \frac{1}{r_1} - \frac{1}{r}, \text{ and } \dot{s} = \omega \frac{rr_1}{r - r_1}.$$

CURVATURE OF ROULETTE.

We may derive some important consequences from the expression just obtained for the the acceleration of a point in the moving plane, namely

$$\ddot{p} = (i\dot{\omega} - \omega^2) \, cp - i\omega \cdot \dot{c}.$$

This consists of three parts; $\omega^2 \cdot pc$ is the acceleration towards c due to rotation about it as a fixed point; $i\dot{\omega} \cdot cp$ is in the direction np perpendicular to cp, due to the change in the angular velocity; and $-i\omega \cdot \dot{c}$ is in the

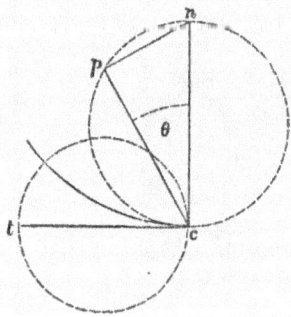

direction cn, due to the change in position of c as the centrode rolls. Hence the normal acceleration of p, that is, the component along pc, is in magnitude $\omega^2 \cdot pc - \omega \cdot \dot{c} \cos\theta$. It vanishes for those points p for which $\omega \cdot pc = \dot{c} \cos\theta$, or for which $cn = \dot{c} : \omega$. These points lie on a circle having cn for diameter; the curvature of this circle is $1 : \frac{1}{2}cn$ or $2\omega : \dot{c}$, that is, it is *twice the difference between the curvatures of the centrodes.* All the points of this circle, therefore, are at the given instant passing through points of inflexion on their paths.

The path of any point p is called a *roulette,* as being traced by rolling motion. We can now determine the curvature of a roulette at any point. For since the normal acceleration is the squared velocity multiplied by the curvature, we have

$$\text{curvature of path of } p = \frac{\omega^2 \cdot pc - \omega \cdot \dot{c}\cos\theta}{\omega^2 \cdot pc^2}$$

$$= \frac{1}{pc} - \frac{\cos\theta}{pc^3} \cdot \frac{rr_1}{r - r_1},$$

where r, r_1 are the radii of curvature of the fixed and rolling centrodes.

The tangential acceleration of p is $\dot{\omega} \cdot cp - \omega\dot{c}\sin\theta$. If therefore we make $ct = \omega\dot{c} : \dot{\omega}$, the locus of points whose tangential acceleration is zero is a circle on ct as diameter. The point at c belongs to both circles; it is a cusp on its path, being a point where there is no normal acceleration, but also no velocity. It has, however, as we know, a tangential acceleration $-i\omega\dot{c}$. The other intersection of the two circles has no acceleration at all.

INSTANTANEOUS AXIS.

In the case of a body moving with one point fixed, we may combine with its velocity-system a spin about any axis through the point, such that the velocity of a certain point a due to the spin is equal and opposite to its actual velocity in the motion of the body. The resultant velocity-system is consistent with rigidity, and the point a is at rest; it is therefore a spin about the axis oa. Consequently the actual motion of the body is a spin about some axis in the plane of these two.

Let oc, $= \omega$, be the instantaneous spin in magnitude and direction, op, $= \rho$, the position vector of any point p. Then we know that the velocity of p is the moment of oc about p, that is, twice the area of the triangle ocp. This quantity, which is in magnitude $oc \cdot op \sin cop$, and in direction perpendicular to oc and op, is what we have called the *vector product* of oc and op, and denoted by $V\omega\rho$. We have therefore

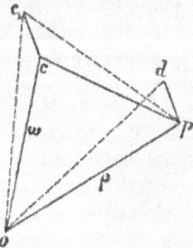

$$\dot{\rho} = V\omega\rho.$$

To find the acceleration of p, therefore, is to find the flux of the triangle ocp, due to the motion of p and c. Now suppose that c moves to c_1 in a certain interval; then $oc_1 p = ocp + coc_1 + c_1 pc$, all the areas being of course regarded as vectors. But if we draw pd equal and parallel to cc_1, we shall have $coc_1 + c_1 pc = pod$, for the three triangles stand on the same base cc_1 or pd, and the height op is the sum of oc and cp. It follows that the flux of ocp, due to the motion of c, is equal to the moment about o of the velocity of c supposed to be transferred to p. That is, the flux of $V\omega\rho$, due to the change of ω, is $V\dot{\omega}\rho$. In a similar way it may be shewn that the flux due to change of ρ is $V\omega\dot{\rho}$. Hence[1] altogether, since $\rho = V\omega\rho$, we have

$$\ddot{\rho} = V\dot{\omega}\rho + V\omega\dot{\rho} = V\dot{\omega}\rho + V \cdot \omega V\omega\rho.$$

The expression $V \cdot \omega V\omega\rho$ means the vector product of the two vectors, ω and $V\omega\rho$. Thus it appears that the acceleration of p consists of two parts; $V \cdot \omega V\omega\rho$ along the perpendicular from p to the axis oc, due to the rotation ω; and $V\dot{\omega}\rho$, perpendicular to op and to the velocity of c, due to the change of ω.

Let a be the point of the moving body at which c is instantaneously situated, then $\ddot{a} = V\dot{\omega}\omega$, or the acceleration of a is equal to the moment about o of the velocity of c.

[1] The flux of a vector product has been already found by a different method on p. 97.

If we interchange the fixed and moving axodes, keeping the relative motion the same, we alter the signs of $\ddot{u}$ and ω; therefore $\dot{\omega}$ is unaltered, or the velocity of c is the same on either supposition. Hence it follows, as in the case of the plane, that the moving axode rolls in contact with the fixed one.

In the general motion of a rigid body, combine with its velocity-system a translation equal and opposite to the velocity of any point a. Then the new velocity-system is a spin round some axis through a. Hence the actual motion is the resultant of a spin and a translation, that is to say, a twist.

DEGREES OF FREEDOM.

The special problems presented by the motion of a plane on a plane are of two kinds. In the first kind, the motion being determined in any way, it is required to find the centrodes. In the second kind, the centrodes being given, it is required to find the path of any point or the envelop of any line in the moving body.

The motion of a plane is determined when each of two curves in the moving plane is made always to touch one of two curves in the fixed plane. Thus the figure bounded by the two curves a, b can be made to move about so that a shall always touch the curve A, and b shall always touch the curve B; and it is clear that its motion is then determined, except as to the time in 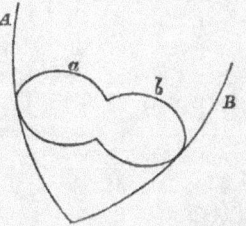 which it is performed. In particular cases one of the curves A, a may shrink into a point; the condition of tangency then resolves itself into the condition that a point in the moving plane shall lie on a fixed curve, or a curve in the moving plane shall pass through a fixed point.

Since it requires three conditions to fix a plane figure in its plane, it is said to have three *degrees of freedom*. If it is subjected to one condition, e. g. that a certain curve must always touch a fixed curve, it has two degrees of freedom left. When it is subject to two conditions, it has one degree of freedom left, and can only move in a certain definite manner.

When one curve has to touch another, the instantaneous centre is situated on the common normal, since the point of contact can only move along the tangent. And as a particular case, when a point has to lie on a given curve, or a curve has to pass through a given point, the instantaneous centre lies on the normal to the curve at that point. In general, if we know the direction of motion of any point, the instantaneous centre is in the line through the point perpendicular to that direction.

INVOLUTE AND EVOLUTE.

For example, if two lines at right angles *pt*, *pn* are made to move as a rigid body, so that *pt* is tangent and *pn* normal to a given curve, the motion of *p* will always be in the direction *pt*, and therefore the instantaneous centre will always be in *pn*. Hence *pn* is the moving centrode; and the fixed centrode, which *pn* rolls upon, is called the *evolute* of the given curve. If *a* is a point where the evolute meets the curve, *pn* = arc *an* in length. The curvature at *p* is 1 : *pn*, by the formula already obtained; thus *n* is the centre of curvature, and the evolute may be described as the locus of centres of curvature of the given curve. Moreover, since *pn* = arc *an*, the curve *ap* may be described by unwinding a string from the curve *an*. On this account *ap* is called an *involute* of the curve *an*. It is clear that every other centre remains in the line *ab*. It now has only one degree of freedom, and the line *ab* is the fixed centrode. The rolling centrode is a curve in the moving plane which shall be called the curve *A*. This curve is clearly the envelop of the line of centres in the moving plane.

Let us now fix the moving plane, and move the fixed plane, subject to the same condition of relative motion. Then as before, for each position there will be a line of centres, and by restricting the instantaneous centre to this, we shall make the motion such that a curve *B* in the plane formerly fixed will roll upon the line of centres. This curve is the envelop of the line of centres in the fixed plane.

Hence the relative motion of the two planes is such that the curves *A* and *B* roll on the same straight line. Or *when a plane slides on a fixed plane, having two degrees of freedom, its motion is such that a curve A in the moving plane rolls on a straight line which rolls on a curve B in the fixed plane.*

Let *x* be any point on the line of centres, and draw the involutes of *A* and *B* which pass through *x*. Then they

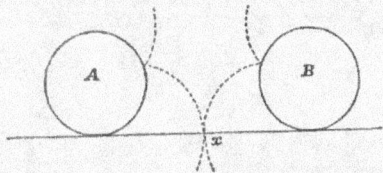

will cut the line at right angles and therefore touch one another at *x*. But if we make *A* roll on the line, having its involute fixed to it, this involute will always pass through *x* at right angles to the line; and similarly for *B*. Hence the relative motion of the two planes is such that these two involutes always touch. Thus *the motion is such that a curve in the moving plane always touches a curve in the fixed plane; but we may substitute for these two curves any two curves parallel to them at equal distances on the same side.*

CHAPTER III. SPECIAL PROBLEMS.

THREE-BAR MOTION.

IF three bars, *ab, bc, cd* are jointed together at *b, c,* while the remaining ends are fixed at points *a, d* about which the bars are free to turn, a plane rigidly attached to

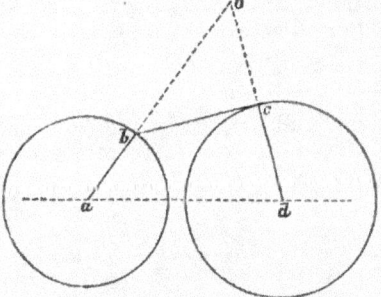

bc is said to have *three-bar motion*. Properly speaking, we ought to consider the jointed quadrilateral *abcd,* and study the relative motion of two of its opposite sides.

We may also specify the motion by saying that the points *b, c* in the moving plane have to lie respectively on two circles in the fixed plane, viz. the circles whose centres are *a, d,* and radii *ab, dc.* The instantaneous centre *o* is at the intersection of *ab* and *dc,* since the motions of *b* and *c* are respectively perpendicular to those lines.

The centrodes of the three-bar motion have only been determined in particular cases. The most important of these is that of the *crossed rhomboid,* so called because its opposite sides are equal. The figure is symmetrical; and if the intersection of ab, cd is at o, we have

$$ao + do = ao + ob = ab \,;$$

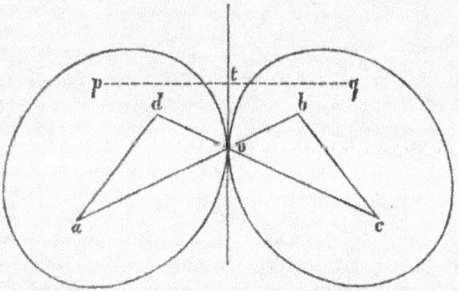

thus the point o describes relatively to ad an ellipse of which a, d are foci, and ab the major axis. Similarly we have $bo + co = ba$, or the locus of o in the moving plane is an equal and similar ellipse. These, therefore, are the centrodes. The relative motion is most clearly understood by supposing both ellipses to roll on the common tangent ot, so as to preserve the symmetrical aspect.

In this way we may see that the path of any point in the moving plane is similar to a pedal of the fixed ellipse. For let p, q be corresponding points in the two ellipses, then the line pq is always bisected at right angles by the tangent ot, and therefore the locus of q, when p is fixed, is similar to the locus of t, but of double the size. It has been proved that the reciprocal of a conic section is always a conic section; from which it follows that the pedal of a conic is also the inverse of a conic (generally a different one; but the same in the case of an equilateral hyperbola in regard to its centre). Hence we see that *every point in the moving plane describes the inverse of a conic.* The inverse of a hyperbola passes twice through the centre of inversion, since the hyperbola goes away to infinity in two directions; but the inverse of an ellipse does not. Hence if q is *outside* the ellipse, so that it can coincide with p in some position of the two curves, it describes the inverse of a hyperbola; but if q is *inside*

10—2

the ellipse, so that it can never reach p, it describes the inverse of an ellipse. Intermediate between these is the case in which q is *on* the ellipse, when the curve which it describes has a cusp and is the inverse of a parabola, which only goes to infinity in one direction.

We have here considered the relative motion of the two *short* sides of a crossed rhomboid. That of the two long sides is equivalent to the rolling of two equal and similar hyperbolas. For in this case we have

$$ao - do = ao - bo = ab,$$

so that the locus of o is a hyperbola having a, d for foci and ab for transverse axis. Remarks may be made about the path of a point in the moving plane entirely similar to those made on the other case.

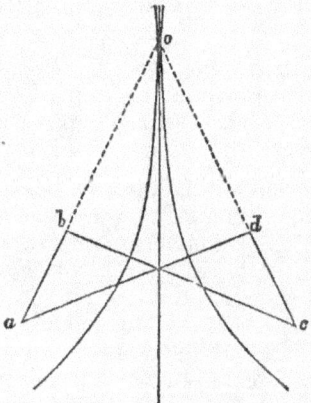

In the general case of three-bar motion, the lengths of the three bars being arbitrary, an important theorem has been obtained by Mr S. Roberts. *Any path described by a point in a plane moving with three-bar motion may also be described in two other ways by three-bar motion.* Suppose (second figure) that ah, hk, kb are the three bars, o the moving point which is rigidly connected with hk by the triangle ohk. Then the theorem is that the path of o may also be described by means of the bars *ag*, *gf*, *fc*, or the bars *bd*, *de*, *ec*. The triangles *hko*, *gof*, *ode* are similar to

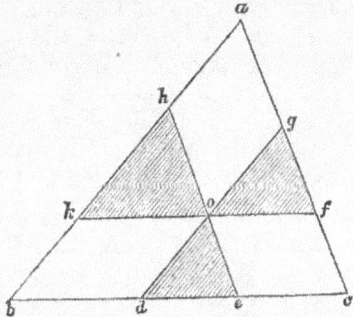

one another, and the figures *ahog*, *bdok*, *cfoe* are parallelograms.

The theorem has been put by Prof. Cayley into the following elegant form. Take any triangle *abc* (first figure) and through any point *o* within it draw lines *kf*, *eh*, *gd* parallel to the sides. Let the triangles *hko*, *gof*, *ode* be

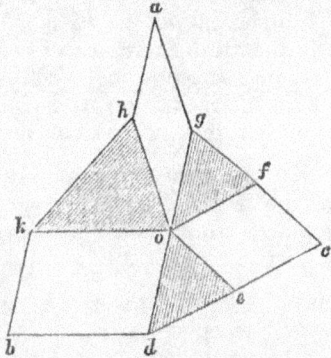

supposed rigid and jointed together at *o*, and let the other lines in the figure represent bars forming three jointed parallelograms. Then *however the system is moved about in its plane* (e.g. into the configuration of the second figure) *the triangle abc will be always of the same shape*. Now the system is one which in shape (independently of its position) has two degrees of freedom; for if we fix one of the three triangles, the other two may be turned round independently. If therefore we impose a single condition, that the area abc shall be constant, the system will still have one degree of freedom. But this is equivalent to fixing the size of abc as well as its shape, so that we may fix the points a, b, c; and still o will be able to move. In so moving it will describe a path which is due at the same time to three different three-bar motions.

All that remains to be proved, therefore, is that the shape of abc is invariable. This can be made clear by very simple considerations. Let q be the operation (complex number) which converts hk into ho, so that $ho = q \cdot hk$. Then the same operation will convert go into gf and od into oe, since the three triangles are similar. Consequently

$$ac = ag + gf + fc = ho + gf + oe = q \cdot hk + q \cdot go$$
$$+ q \cdot od = q(hk + ah + kb) = q \cdot ab,$$

that is, ac is got from ab by the same operation which converts hk into ho; therefore the triangle abc is similar to hko. Or in words, the components ag, gf, fc of ac are got from the components hk, ah, kb of ab by altering all their lengths in the same constant ratio and turning them all through the same constant angle. Therefore the whole step ac is got from ab by altering its length in a constant ratio and turning it through a constant angle.

It is to be observed that the configuration in the first figure forms an apparent exception to the theorem. The area abc is then a maximum, and the path of o has shrunk up into a point, so that it is really not able to move.

We may use Mr Roberts' theorem to transform motion due to the crossed rhomboid into that due to a figure called a *kite* by Prof. Sylvester. It also is a quadrilateral having its sides equal two and two, but the equal sides are adjacent.

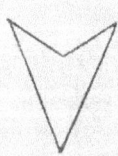

Now if the point o be taken in the first figure so that gd is bisected at o, the triangles gof, ode will be equal in

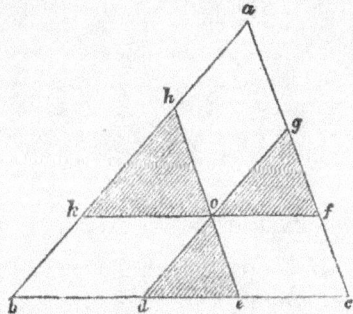

all respects, and bk will equal ha. Now put the figure into such a configuration that abc is equal to hko: then $ahkb$ is a crossed rhomboid, and both the figures $bdec$, $agfc$ are kites. For $\qquad bc = ko = bd$, and $ac = ho = ag$,

while $de = ec$ and $gf = fc$ by construction.

It follows that in the three-bar motion determined by a kite, the path of every point in the moving plane is the inverse of a conic; since it may also be described by means of a crossed rhomboid.

CIRCULAR ROULETTES.

Considerable interest attaches to the case of plane motion in which both centrodes are circles, or when one is a circle and the other a straight line; the latter being a speciality of the former, obtained by making the radius of one circle infinite. The path traced by a point in the circumference of the rolling circle is called a cycloidal curve, that traced by any other point in the moving plane a trochoidal curve; the names cycloid and trochoid *simpliciter* being applied to paths traced in the rolling of a circle on a straight line.

Two circles may touch each other so that each is outside the other, or so that one includes the other. In the former case, if one circle rolls on the other, the curves traced are called epicycloids and epitrochoids. In the latter case, if the inner circle roll on the outer, the curves are hypocycloids and hypotrochoids; but if the outer circle roll on the inner, the curves are epicycloids and peritrochoids. We do not want the name pericycloids, because, as will be seen, every pericycloid is also an epicycloid; but there are three distinct kinds of trochoidal curves.

DOUBLE GENERATION OF CYCLOIDAL CURVES.

Every cycloidal curve (except the cycloid *par excellence*) can be *generated in two different ways.* In the case of hypocycloids, let a and b be centres of two circles the *sum* of whose radii is equal to the radius of the fixed circle. Then if we complete the parallelogram $oapb$, p will be a point of intersection of these circles, for $ap = ob = od - bd =$ radius of circle a, and similarly bp equal radius of circle b. Hence the

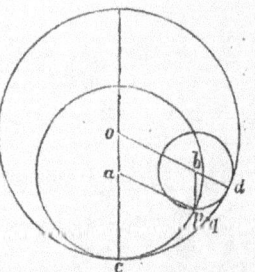

angles cap, pbd, cod are all equal, and therefore the arcs ap, pd, ad are the same portion of the circumferences of their several circles. But the radius of the large circle is the sum of the radii of the two smaller ones; therefore its circumference is the sum of their circumferences, and consequently the arc cd is the sum of the arcs cp, pd. Make $cq = cp$, so that $qd = pd$; then by the rolling of the circle a the point p would come to q, and by the rolling of the circle b the point p would come to q; hence the intersec-

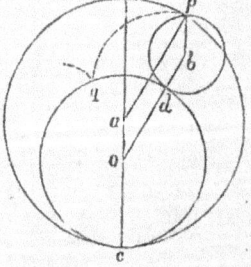

tion of the two circles is a fixed point on each of them, and the path of p may be described by the rolling of either.

In the case of epicycloids, the *difference* of radii of the rolling circles is equal to the radius of the fixed circle; the arc cp is equal to $cd + dp$, and p would be brought to q by the rolling of either a or b.

CASE OF RADII AS 1 : 2.

A very important case is that of internal rolling, the radius of one circle being half that of the other. Draw the straight line opq to meet both circles. Let $cop = \theta$,

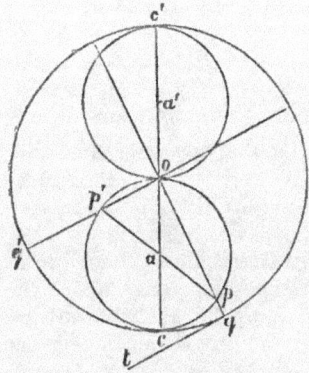

then $cap = 2\theta$; and if a be the radius of the smaller circle, $2a$ of the larger, arc $cp = 2a\theta$, and arc $cq = 2a\theta$; therefore $cp = cq$, or p will come to q in the rolling. Hence every point in the circumference of the rolling circle describes a diameter of the fixed circle. The opposite point p' describes the diameter perpendicular to the former.

If we suppose the small circle to roll from c to the right, oa will turn counterclockwise into the position oa', while ac will turn clockwise into the position $a'c'$. Hence the motion (supposing the rolling to take place uniformly) is a composition of two circular motions of the *same period* in opposite directions. Consequently the motion of any point can be resolved into simple harmonic motions all of the same period. It follows that the motion of every point of the moving plane is harmonic motion in an ellipse, which in certain cases as we have seen reduces itself to simple harmonic motion on a diameter of the fixed circle.

Hence if a line of constant length ab be moved with its extremities on two fixed lines oX, oY, the path of every point rigidly connected with ab will be an ellipse with centre o, unless the point is on the circumference of the circle circumscribing oab, in which case the path is a straight line through o. An apparatus for describing an ellipse by means of

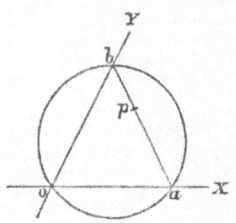

a pencil attached at a point p of a bar so moving, is called the *elliptic compasses*. The semi-diameters of the ellipse along oX and oY are pa, pb respectively. When oX, oY are at right angles, these are the semi-axes of the ellipse.

If the small circle be fixed, and the larger roll round it, the motion is such that every diameter of the rolling circle passes through a fixed point on the small one. Now every line in the moving plane is parallel to some diameter of the large circle, and must therefore remain at a fixed distance from the point through which the diameter always passes; consequently it always touches a circle whose centre is at that point. Hence *every straight line in the moving plane envelops a circle*. Conversely, if a plane move so that two straight lines in it always touch two fixed circles, then every line in the plane will envelop a circle. For two lines parallel to them through the centres of the circles are fixed relatively to the moving plane; thus a line of constant length in the fixed plane always has its extremities on two lines of the moving plane, and the motion is the one here considered.

The curve traced by a point in the circumference of the large circle is a *cardioid*, which we have already met with as the inverse of a parabola in regard to the focus, or, which is the same thing, the pedal of a circle in regard to a point on the circumference. If the point q describe a cardioid, the line qt, tangent to the large circle, always touches a fixed circle whose centre is at p', and which therefore touches the fixed small circle at p. Hence q is the foot of the perpendicular on the tangent to this circle from the point p on its circumference. The cardioid may also be described by the *external* rolling of a circle on a fixed circle of equal size.

ENVELOP OF CARRIED ROULETTE.

When a circle rolls on a fixed circle, every diameter of the rolling circle envelops a cycloidal curve. Suppose a circle of half the size to roll together with the circle o, so as to have always the same point of contact; then the relative motion of these two circles will be that which we have just considered, and a point p, fixed on the small circle,

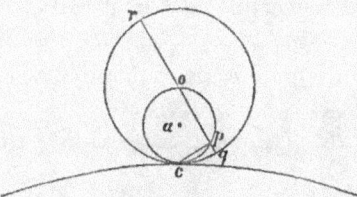

will be always on the *diameter oq*. The tangent to the cycloidal path described by p, in consequence of the rolling of the circle a on the fixed circle, is opq, since c is the instantaneous centre ; hence this line always touches the cycloidal curve.

This theorem is a particular case of the following. Let

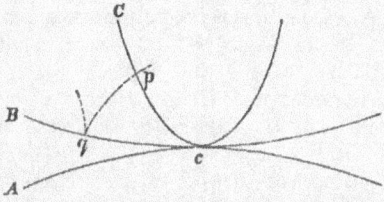

a curve B roll on a curve A, carrying with it the roulette pq made by rolling C on B; then the envelop of this roulette is a curve which may be described by rolling C on A. Suppose B and C to roll simultaneously on A, so as always to have the same point of contact; then the motion of C relative to B is that which describes the roulette qp. Now cp is perpendicular to the tangent both of this roulette and of that which p describes by the rolling of C on A. Hence the two roulettes always touch one another, as was to be proved. Observe that the point p is not necessarily on the curve C.

Returning to the case of the circles, we observe that the extremities q, r of the moving diameter describe similar and equal cycloidal curves, such that a cusp of one and a vertex of the other are on the same diameter of the fixed circle. Hence if a straight line of constant length move with its ends on two such cycloidal curves, starting from a position in which one end is at a cusp and the other at a corresponding vertex, it will envelop a cycloidal curve.

The following are cases of this theorem:

1. *The chord of a cardioid through the cusp is of constant length.* (A point is a special case of a cycloidal curve.)

2. *A line of constant length with its ends moving in two fixed lines at right angles envelops a four-cusped hypocycloid.*

3. *The portion of the tangent to a three-cusped hypocycloid intercepted by the curve is of constant length.*

The curvature of cycloidal curves may be calculated by means of the general theorem already given for the curvature of roulettes, or directly as follows. Let o be the centre of the fixed circle, take $ce : dc = do : co$, draw a circle through e with centre o, and a circle on ce as diameter. Produce pc to meet this in q. If this circle roll on the circle through e, so that q is brought to h, we shall have $eq = eh$, and since $eq : pd = ec : cd = oe : oc$, pd is equal to the corresponding arc of the circle kc. Hence the two small circles may roll together on the two large ones, so that ce always passes through o, and pcq is a straight line. Then

pq is normal to the path of p and tangent to that of q, or the latter path is the evolute of the former.

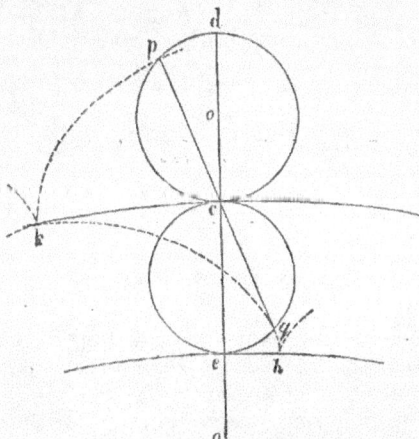

It follows that the length of the arc kq is equal to pq, or $s = de \cos \psi$. It is clear that ψ is in a fixed ratio to the angle ϕ which pq makes with the normal at k, and consequently $s = a \cos m\phi$, if $a = de$ and m is this fixed ratio.

BOOK III. STRAINS.

CHAPTER I. STRAIN-STEPS.

STRAIN IN STRAIGHT LINE.

WE have hitherto studied the motion of *rigid* bodies, which do not change in size or shape. We have now to take account of those *strains*, or changes in size and shape, which we have hitherto neglected.

The simplest kind of strain is the change of length of an elastic string when it is stretched or allowed to contract. When every portion of the string has its length altered in the same ratio, the strain is called *uniform* or *homogeneous*. Thus if apb is changed into $a'p'b'$ by a uniform strain, $ap : a'p' = ab : a'b'$. The ratio 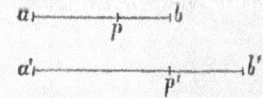 $a'p' : ap$, or the quantity by which the original length must be multiplied to get the new length, is called the *ratio* of the strain. The ratio of the change of length to the original length, or $a'p' - ap : ap$, is called the *elongation*; it is reckoned negative when the length is diminished. A negative elongation is also called a *compression*.

Let e be the elongation, s the unstretched length ap, σ the stretched length $a'p'$, then $\sigma - s = es$, or $\sigma = s\,(1 + e)$. Thus $1 + e$ is the ratio of the strain.

In general, a solid body undergoes a strain of simple elongation e, when all lines parallel to a certain direction are altered in the same ratio $1 : 1 + e$, and no lines perpendicular to them are altered in magnitude or direction.

The strain is then entirely described if we describe the strain of one of the parallel lines.

HOMOGENEOUS STRAIN IN PLANE.

The kind of strain next in simplicity is that of a flat membrane or sheet. Suppose this to be in the shape of a square; we may give it a uniform elongation e parallel to one side, and then another uniform elongation f parallel to the other side. It is now converted into a rectangle, whose sides are proportional to $1 + e$, $1 + f$. By each of these operations two equal and parallel lines, drawn on the membrane, will be left equal and parallel; though, if not parallel to a side of the square, they will be altered in direction.

We may prove, conversely, that every strain which leaves straight lines straight, and parallel lines parallel, is a strain of this kind combined with a change of position of the membrane in its plane. Such a strain is called *uniform* or *homogeneous*.

Since a parallelogram remains a parallelogram, *equal parallel lines remain equal*. Then it is easy to shew, by the method of equi-multiples, that the ratio of *any two parallel lines is unaltered* by the strain. Next, if we draw a circle on the unstrained membrane, this circle will be altered by the strain into an ellipse. For in the unstrained figure

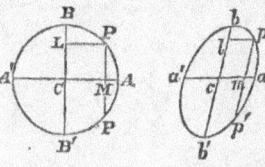

$$A'M \cdot MA : CA^2 = MP^2 : CB^2,$$

and since these ratios of parallel lines are unaltered, it follows that in the strained figure also

$$a'm \cdot ma : ca^2 = mp^2 : cb^2.$$

Hence the strained figure is an ellipse, whose conjugate diameters are the strained positions of perpendicular diameters of the circle.

It follows that there are two directions at right angles to one another, which remain perpendicular after the strain; namely those which become the *axes* of the ellipse into which a circle is converted. If these lines remain parallel to their original directions, the strain is produced by two simple elongations along them respectively; in that case it is called a *pure* strain. If they are not parallel to their original directions, the strain is compounded of a pure strain and a rotation.

Two lines drawn anywhere in the strained membrane parallel to the axes of the ellipse into which a circle is converted, or in the unstrained membrane parallel to the unstrained position of those axes, are called *principal axes* of the strain. The elongations along them are called *principal elongations;* the ratios in which they are altered are called principal ratios.

REPRESENTATION OF PURE STRAIN BY ELLIPSE.

When the strain is pure, the new position of any step may be conveniently represented by means of a certain ellipse. Let the principal ratios be p, q, so that every line parallel to oX is altered in the ratio $1 : p$, and every line parallel to oY in the ratio $1 : q$. Take two lengths oa, ob, along oX, oY respectively, such that $oa^2 : ob^2 = q : p$, and let m be the positive geometric mean of p, q, so that $m^2 = pq$. Then we shall have, so far as length is concerned, $p \cdot oa = m \cdot ob$, and $q \cdot ob = m \cdot oa$. Hence, taking account of direction, oa becomes $im \cdot ob'$, and ob becomes $im \cdot oa$, in consequence of the strain.

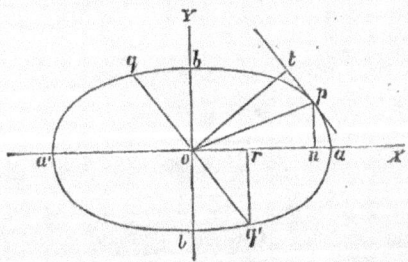

Now construct an ellipse having oa, ob for semi-axes; then if p be any point on it and qq' the diameter conjugate to op, the strain will turn op into $im . oq'$. For since it turns oa into $im . ob'$, it will turn on into $im . rq'$, because $on : oa = rq' : ob'$ (p. 129). And since it turns ob into $im . oa$, it will turn np into $im . or$, because $np : ob = or : oa$. Therefore it will turn op, which is $on + np$, into $im (rq' + or)$, that is, into $im . oq'$.

Hence we see that the strained position of any vector is perpendicular to the conjugate diameter of a certain ellipse, having that vector as diameter, and is proportional to the conjugate diameter in length. For the ellipse used in this representation may be of any size, since all that is necessary for it is that its axes should be parallel to the principal axes of the strain, and inversely proportional to the square roots of the principal ratios.

REPRESENTATION OF THE DISPLACEMENT.

The *displacement* of any point is the step from its old position to its new one. Thus if a vector op is turned by the strain into op', the displacement of p is pp'.

When the two principal elongations e, f are of the same sign, the displacement may be represented by an ellipse, in the same way as we have represented the new position of any vector. The only difference is that we are now to draw an ellipse whose axes are inversely proportional to the square roots of the *elongations*, so that $oa^2 : ob^2 = f : e$, and to make $m^2 = ef$, giving to m the same sign as e or f. Then the displacement of a will be $im . ob'$, and the displacement of b will be $im . oa$. Hence it follows (as before) that the displacement of p will be $im . oq'$. In this case therefore the displacement of every point on the ellipse is perpendicular and proportional to that diameter which is parallel to the tangent at the point.

But when e and f are of different signs, it is necessary to use a hyperbola to represent the displacement. Let $m^2 = - ef$, and $oa^2 : ob^2 = - f : e$; and let m be taken of the same sign as f. Then the displacement of a will be $im . ob$, and the displacement of b will be $im . oa$. If then a hyperbola be described with oa and ob as axes, and *op*, *oq* be a pair of conjugate semi-diameters, the displacement of *p* will be *im* . *oq* and that of *q* will be *im* . *op*. The

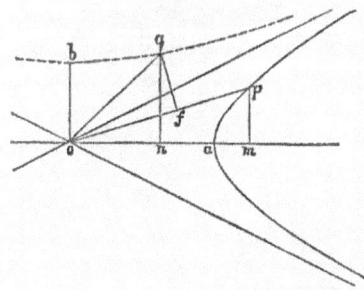

proof is the same as for the ellipse, depending on the property that *np* : *ob* = *or* : *oa*, and *rq* : *ob* = *on* : *oa*.

The ellipse or hyperbola which is thus used to represent the displacement is called the *displacement-conic* of the strain.

LINEAR FUNCTION OF A VECTOR.

One vector is said to be a *function* of another, when its components are functions of the components of the other ; so that, for every value (including magnitude and direction) of one of them, there is a value or values of the other. Thus $pi + qj + rk$ is a function of $xi + yj + zk$ if p, q, r are functions of x, y, z. We may express this relation between them thus: $pi + qj + rk = \phi (xi + yj + zk)$.

A function of a vector is said to be *linear* when that function of the sum of two vectors is the sum of the functions of the vectors. Thus the function ϕ is linear when

$$\phi (\alpha + \beta) = \phi \alpha + \phi \beta.$$

At present we shall consider only linear functions of vectors which are all in one plane. It is clear that when a plane figure receives a homogeneous strain, the strained position of any vector is a linear function of the vector. For the triangle made of two vectors α, β and their sum $\alpha + \beta$ becomes after the strain a triangle made of the vectors $\phi\alpha$, $\phi\beta$, $\phi (\alpha + \beta)$;
and consequently $\phi (\alpha + \beta) = \phi \alpha + \phi \beta$.

If the strain is not homogeneous, the strained position of any finite line pq is not a function merely of the vector pq, that is, of the length and direction of the line, but also of its position.

The displacement of any point is also a linear function of its position-vector in regard to an origin supposed to remain fixed during the strain. For let α, β be two sides of a parallelogram, one corner of which is the origin and the opposite corner of which is consequently the point whose position-vector is $\alpha + \beta$; then since this parallelogram remains a parallelogram during the strain, the displacement of the corner $\alpha + \beta$ is the resultant of the displacements of the corners α, β. Hence if $\psi\alpha$ is the displacement of the end of α,

$$\psi\,(\alpha + \beta) = \psi\alpha + \psi\beta.$$

If $\phi\alpha$ is the strained position of α, the displacement of its end is $\phi\alpha - \alpha$. Hence the *strain-function* ϕ and the *displacement-function* ψ are connected by the equation,

$$\phi\alpha = \psi\alpha + \alpha, \quad \text{or} \quad \phi\alpha = (\psi + 1)\alpha,$$

which may be written $\phi = \psi + 1$.

If n is a number or scalar quantity, $\phi\,(n\alpha) = n\phi\alpha$ when the function ϕ is linear; for since functions of equal lengths measured in the same direction are equal, and functions of multiples of such lengths are multiples of the functions of the lengths, it follows that functions of unequal lengths are proportional to those lengths. Hence it follows that

$$\phi(xi + yj) = x\phi i + y\phi j,$$

and therefore we know the function of every vector in the plane when we know the functions of i and j. Let then $\phi i = ai + hj$, $\phi j = h'i + bj$; then the function ϕ is completely known when the four quantities a, h, h', b are known. These equations may be abbreviated into the form

$$\phi i,\ \phi j = (a,\ h)(i, j) \quad \text{or} \quad \phi = (a,\ h)$$
$$\qquad\qquad |h',\ b| \qquad\qquad\qquad |h'\ b|$$

The set of four quantities a, h, h', b, written in a square shape as in the last formula, is called a *matrix*; thus we

11—2

may say that the function ϕ is determined by its matrix. The matrix must be carefully distinguished from its *determinant*, which is the single quantity $ab - hh'$, calculated from the four constituents a, h, h', b of the matrix.

PROPERTIES OF A PURE FUNCTION.

The strain-function and the displacement-function of a pure strain are both called *pure* functions. We proceed to investigate what must be the relation between the quantities a, h, h', b in order that the function ϕ may be pure.

If ϕ is the strain-function of a pure strain, $Sp\phi\sigma = S\sigma\phi\rho$, where ρ, σ are any two vectors. Let op and or be semi-diameters parallel to ρ, σ, of the ellipse which represents the strain; then if oq and os are the conjugate semi-diameters, $\phi(op) = im \cdot oq$ and $\phi(or) = im \cdot os$.

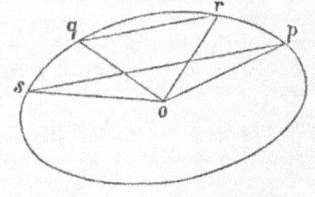

Thus the *cosine* of the angle which op makes with $\phi(or)$ is the *sine* of the angle it makes with os. Therefore the scalar product of op and $\phi(or)$ is twice the triangle ops, and the scalar product of or and $\phi(op)$ is twice the triangle orq. But that these triangles are equal appears at once from intuition of the corresponding figure in the circle of which the ellipse is orthogonal projection ; where the angles POQ, ROS will be right angles. Therefore $S \cdot op \cdot \phi(oq) = S \cdot oq \cdot \phi(op)$; let then $\rho = x \cdot op$, $\sigma = y \cdot oq$, and $Sp\phi\sigma$ will be

$$xy\, S \cdot op \cdot \phi(oq), = xy\, S \cdot oq \cdot \phi(op) = S\sigma\phi\rho.$$

It follows immediately that the same property belongs to the displacement-function. For let $\phi\rho = \rho + \psi\rho$, so that ψ is the displacement-function of the strain ϕ. Then we have

$$Sp(\sigma + \psi\sigma) = Sp\phi\sigma = S\sigma\phi\rho = S\sigma(\rho + \psi\rho)$$

and therefore, since $S\sigma\sigma = S\sigma\rho$,

$$Sp\psi\sigma = S\sigma\psi\rho.$$

Now although the strained position of a vector, in any actual strain, is represented by an ellipse in the manner just made use of; the *displacement* may be represented by a hyperbola, and the equation

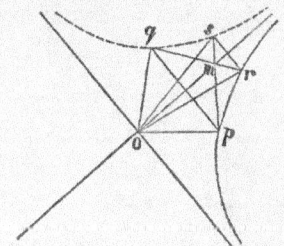

$$Sp\psi\sigma = So\psi\rho$$

will involve in that case an equality of triangles like that which has been just proved for the ellipse. The theorem is however obvious; since the parallel lines pq, rs are bisected by the asymptote, it goes through the intersection m of ps, qr; therefore the triangles omp, omr are respectively equal to oqm, osm, and therefore by addition $ops = orq$.

Suppose, as before, that

$$\phi i = ai + hj, = op, \quad \phi j = h'i + bj, = oq.$$

Then $\qquad a = om, \ h = mp, \ h' = on, \ b = nq.$

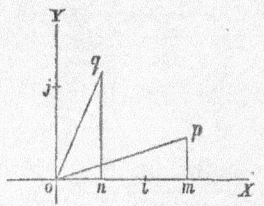

The *magnitude* of $Si\phi j$, the scalar product of oi and oq, is the product of the length of oi by the length of the component of oq along it; that is, it is $oi \cdot on$ or h', since the length of oi is unity. (For reasons to be subsequently explained, the scalar product of two vectors is taken to be the *negative* product of either by the component of the other along it; this is a convention, and does not affect the present argument.) Hence we have $Si\phi j = -h'$; similarly $Sj\phi i = -h$. If the function ϕ is pure, $Si\phi j = Sj\phi i$; thus *for the function ϕ to be pure, it is necessary that $h = h'$.*

To shew conversely that when $h = h'$ the function is pure, we shall actually find the principal axes and elongations of the strain of which it is the displacement-function. Let e, f be the principal elongations, θ the angle between oX and the axis of elongation e. A step of unit length making the angle θ with oX is $i \cos \theta + j \sin \theta$, and a unit step at right angles to this is $i \sin \theta - j \cos \theta$. One of these receives the elongation e and the other the elongation f, each in its own direction; therefore

$$\phi (i \cos \theta + j \sin \theta) = e(i \cos \theta + j \sin \theta),$$
$$\phi (i \sin \theta - j \cos \theta) = f(i \sin \theta - j \cos \theta).$$

Multiply the first equation by $\cos \theta$, the second by $\sin \theta$, and add; thus we get

$$\phi i = (e \cos^2 \theta + f \sin^2 \theta) \, i + (e - f) \sin \theta \cos \theta \, . \, j = ai + hj.$$

Similarly, by multiplying the first equation by $\sin \theta$, the second by $\cos \theta$, and subtracting, we get

$$\phi j = (e - f) \sin \theta \cos \theta \, . \, i + (e \sin^2 \theta + f \cos^2 \theta) \, . \, j = hi + bj.$$

It is now necessary to find quantities, e, f, θ, which satisfy the equations

$$e \cos^2 \theta + f \sin^2 \theta = a,$$
$$e \sin^2 \theta + f \cos^2 \theta = b,$$
$$(e - f) \sin \theta \cos \theta, = \tfrac{1}{2} (e - f) \sin 2\theta = h.$$

Adding the first two, we have

$$e + f = a + b;$$

subtracting the second from the first,

$$(e - f) \cos 2\theta = a - b;$$

combining this with the third,

$$(e - f)^2 = 4h^2 + (a - b)^2.$$

Consequently

$$\tan \theta = \frac{2h}{a - b},$$
$$2e = a + b + \sqrt{\{4h^2 + (a - b)^2\}},$$
$$2f = a + b - \sqrt{\{4h^2 + (a - b)^2\}}.$$

Compare with this the solution of an analogous problem on p. 131, making in that, $\theta = \tfrac{1}{2}\pi$, and $k_1 = k_2$.

SHEAR.

When the plane is as much lengthened along one principal axis of the strain as it is shortened along the other, so that $(1 + e)(1 + f) = 1$, or $e + f + ef = 0$, the strain is called a *shear*. In this case it is clear that the area of every figure in the plane remains unaltered.

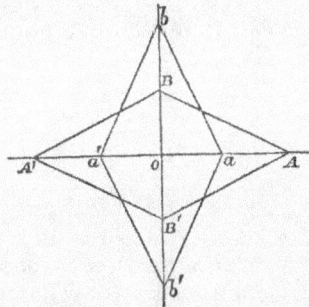

Let oa be changed into oA, and take $ob = oA$; then ob will be changed into oB, which is equal to oa. Hence the rhomb $aba'b'$ will become the rhomb $ABA'B'$, and ab, which becomes AB, will be unaltered in length. If we combine this pure strain with a rotation, so as to bring ab

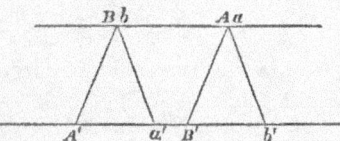

to coincide with AB, then $a'b'$ may be brought to $A'B'$ by a sliding motion along its line. Thus all lines parallel to ab will be slid along themselves through lengths proportional to their distances from ab. The amount of sliding per unit distance is called the *amount* of the shear.

Since we have also $a'b = A'B$, the shear might also be produced by the sliding of lines parallel to $a'b$; but then it would be combined with a different rotation. Thus there are two sets of parallel lines which are unaltered in length, and whose relative motion is a sliding along themselves.

The ratio $oA : oa$ is called the *ratio* of the shear. If $ob = \alpha \cdot oa$, the sliding of $a'b$ relative to $b'a$ is $2ab \cdot \cos aba'$ and the distance between $a'b$ and $b'a$ is $ab \sin aba'$. Hence the amount of the shear is $2 \cot aba' = 2 \cot 2\theta$, if $\theta = abo$, so that $\cot \theta = \alpha$. Now

$$2 \cot 2\theta = \frac{\cos^2 \theta - \sin^2 \theta}{\sin \theta \cos \theta} = \cot \theta - \tan \theta = \alpha - \frac{1}{\alpha}.$$

Thus, if α be the ratio of a shear, its amount s is given by $s = \alpha - \alpha^{-1}$.

We have seen that e and f satisfy the equation

$$e + f + ef = 0,$$

in the case of a shear. When e and f are very small fractions, ef is small compared with either of them, and we have approximately $e + f = 0$. The ratio $-e : f$ differs from unity, in fact, by the small fraction e. Thus the displacement-conic is approximately a rectangular hyperbola.

Now the ratio of the shear is $1 + e$, and

$$(1 + e)(1 + f) = 1.$$

Hence the amount is

$$1 + e - (1 + f) = e - f ;$$

this is accurate, whether the shear be large or small. But if the shear is very small, f is approximately equal to $-e$, and thus the amount is approximately $= 2e$.

COMPOSITION OF STRAINS.

When the displacement of every point, due to a certain strain, is the resultant of its displacements due to two or more other strains, the first strain is said to be the *resultant* of these latter, which are called its *components*. If the displacement of the end of ρ in two strains respectively be $\phi\rho$ and $\psi\rho$, the displacement in their resultant is $(\phi + \psi)\rho$.

This must be carefully distinguished from the result of making a body undergo the two strains successively. Thus if ρ be changed into $\phi_1\rho$ by the first strain, and into $\psi_1\rho$ by the second, the effect of applying the second strain after the first will be to change ρ into $\psi_1\{\phi_1(\rho)\}$ or $\psi_1\phi_1\rho$. To compare this with the preceding expression for the resultant, we must observe that $\phi_1 = 1 + \phi$ and $\psi_1 = 1 + \psi$; so that whereas in the one case the displacement is $(\phi + \psi)\rho$, in the other it is $(\phi + \psi + \psi\phi)\rho$. In one case only the *addition*, in the other the *multiplication* of functions is involved. For this reason we shall speak of the strain, whose effect is the same as that of two other strains successively applied, as the *product* of the two strains.

A strain in which $a = b = 0$, and $h = -h'$, is called a *skew* strain, and the displacement-function ϕ a skew function. It is the *product* of a rotation about the origin and a uniform dilatation; for the displacement of every point p is perpendicular to op and proportional to it.

Every plane strain is the *resultant* of a pure and a skew strain. For let a, h, h', b have the same meaning as before; these numbers are the sums of

$$a, \tfrac{1}{2}(h + h'), \tfrac{1}{2}(h + h'), b, \text{ and } 0, \tfrac{1}{2}(h - h'), \tfrac{1}{2}(h' - h), 0,$$

of which the former belong to a pure, and the latter to a skew strain.

But every plane strain is the *product* of a rotation, a uniform dilatation, and a shear. First rotate the plane till the principal axes of the strain are brought into position; then give it uniform dilatation (or compression) till the area of any portion is equal to the strained area; the remaining change can be produced by a pure shear.

When two strains are both very small, their product and resultant are approximately the same strain.

REPRESENTATION OF STRAINS BY VECTORS.

We have seen that if e, f be the principal elongations of a pure strain (a, h, h, b), then $e + f = a + b$. Hence if $a + b = o$, we must have $e + f = o$. Hence the strain is made by an elongation in one direction, combined with an equal compression in the perpendicular direction. Such a strain is approximately a shear when it is very small; we shall therefore call it a *wry shear*. Its characteristic is that its displacement-conic is a rectangular hyperbola. A wry shear accompanied by rotation shall be called a *wry strain;* that is (a, h, h', b) is a wry strain if $a + b = 0$.

Every strain is the resultant of a uniform dilatation and a wry strain. For

$$(a, h, h', b) = \tfrac{1}{2}(a + b, 0, 0, a + b) + \tfrac{1}{2}(a - b, 2h, 2h', b - a).$$

Every wry strain is the resultant of a skew strain and a wry shear. For

$$\tfrac{1}{2}(a - b, 2h, 2h', b - a) = \tfrac{1}{2}(0, h - h', h' - h, 0)$$
$$+ \tfrac{1}{2}(a - b, h + h', h + h', b - a).$$

The *magnitude* of a skew strain $(0, h, -h, 0)$ is h. Being the product of a rotation by a uniform dilatation, it is not specially related to any direction in the plane, and may therefore be represented by a vector of length h perpendicular to the plane.

The wry shear $(a, h, h, -a)$ has for its displacement-conic a rectangular hyperbola whose transverse axis makes with oX an angle θ such that $\tan 2\theta = h : a$ (since in this case $a - b = 2a$; the general value being $\tan \theta = 2h : a - b$). Moreover if $e, -e$ are its principal elongations, we have in general $(e - f)^2 = (a - b)^2 + 4h^2$, and therefore in this case $e^2 = a^2 + h^2$. Hence if a wry shear be represented by a vector in its plane, of length equal to its positive principal elongation, making with oX an angle (2θ) equal to *twice* the angle (θ) which that elongation makes with it; the components (a, h) of this vector along oX and oY will represent in the same way the wry shears $(a, 0, 0, -a)$ and

$(0, h, h, 0)$, having oX and oY respectively for axes and asymptotes, of which the given wry shear is the resultant. Let such a vector be called the *base* of the wry shear; then our proposition is that *the base of the resultant of two wry shears is the resultant of their bases.*

This is obvious, because the base of $(a, h, h, -a)$ is $ai + hj$.

This mode of representation is to a certain extent arbitrary, because it depends upon the position of oX. It will, however, be found useful in many ways.

Combining this with our previous representation of a skew strain, we see that a wry strain in general may be represented by a vector not necessarily in its plane, the normal component of which represents the skew part of the strain, while the component in the plane represents the wry shear.

When a figure receives a uniform dilatation, without rotation, we may regard it as merely multiplied by a numerical ratio or scalar quantity. Thus the whole operation of any plane strain may be regarded as the sum of a scalar and a vector part. If we write, for example,

$$1 = (1, 0, \quad 0, \quad 1) \ldots \text{(leaves the figure unaltered)}$$

$$I = (1, 0, \quad 0, -1) \ldots \text{(turns it over about } oY\text{)}$$

$$J = (0, 1, \quad 1, \quad 0) \ldots \text{(interchanges } oX \text{ and } oY\text{)}$$

$$K = (0, 1, -1, \quad 0) \ldots \text{(turns counter clock-wise through a right angle)}$$

then we shall have

$$(a, h, h', b) = \tfrac{1}{2}(a+b) + \tfrac{1}{2}(a-b)I + \tfrac{1}{2}(h+h')J + \tfrac{1}{2}(h-h')K,$$

and it will be easy, by combining these operations, to verify that $I^2 = 1$, $J^2 = 1$, $K^2 = -1$, $JK = I = -KJ$, $KI = J = -IK$, $IJ = K = -JI$.

GENERAL STRAIN OF SOLID.   PROPERTIES OF THE ELLIPSOID.

When a solid is so strained that the lengths of all parallel lines in it are altered in the same ratio, it is said to undergo *uniform* or *homogeneous* strain.   It follows easily, as before, that all parallel planes remain parallel planes, and undergo the same homogeneous strain, besides being altered in their aspect.

A sphere is changed into a surface which is called an *ellipsoid*, having the property that every plane section of it is an ellipse.   We may easily obtain its principal properties from those of the sphere, if we remember only that the ratios of parallel lines are unaltered by the strain.

Thus we know that if a plane be drawn through the centre of a sphere, the tangent planes at all points where it cuts the sphere are perpendicular to it, and therefore parallel to the normal to it through the centre; this normal meets the sphere in two points where the tangent planes are parallel to the first plane.

A plane A drawn through the *centre* of the ellipsoid (a point such that all chords through it are bisected at it) is called a *diametral* plane.   The tangent planes at all points where it cuts the surface are parallel to a certain line through the centre, called the *diameter conjugate* to the given plane; this line cuts the surface in two points where the tangent planes are parallel to the given plane A.

Any two conjugate diameters of the ellipse in which the ellipsoid is cut by the plane A, together with the diameter conjugate to that plane, form a *system of three conjugate diameters;* each of them is conjugate to the plane containing the other two.   They correspond to three diameters of a sphere at right angles to one another.   The planes containing them two and two are called conjugate diametral planes.

Let oa, ob, oc be three conjugate semi-diameters of the ellipsoid, p any point on the surface; draw pn parallel to

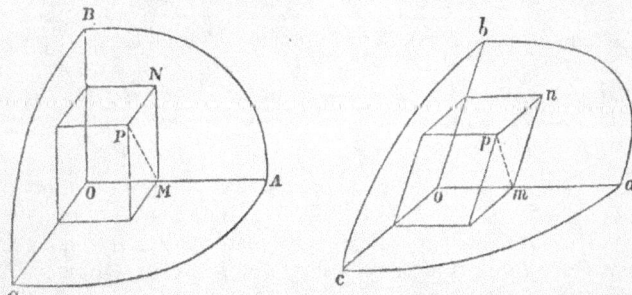

oc to meet the plane oab in n, and then draw nm parallel to ob to meet oa in m. These points will be the strained positions of O, A, B, C, P, M, N, when OA, OB, OC, are at right angles and P is a point on the sphere. Now

$$OP^2 = OM^2 + MP^2 = OM^2 + MN^2 + NP^2;$$

or, remembering that $OA = OB = OC = OP$, we have

$$\frac{OM^2}{OA^2} + \frac{MN^2}{OB^2} + \frac{NP^2}{OC^2} = 1.$$

But the ratios of parallel lines being unaltered by the strain, $OM : OA = om : oa$, and $MN : OB = mn : ob$; hence in the ellipsoid also we have

$$\frac{om^2}{oa^2} + \frac{mn^2}{ob^2} + \frac{np^2}{oc^2} = 1, \qquad \text{or} \quad \frac{x^2}{a^2} + \frac{y^2}{b^2} + \frac{z^2}{c^2} = 1,$$

if x, y, z are written for om, mn, np, and a, b, c for oa, ob, oc.

Let a plane be drawn through O perpendicular to OP. It will cut the sphere in a great circle, whose area shall be called A; the area is of course the same for all sections of the sphere by planes through O. The angle between this plane and OBC will be the same as the angle between OP and OA, the straight lines perpendicular to those planes respectively. Call this angle θ. Then if we project the area A on the plane OBC, the area of the projection will be $A \cos \theta$. Now A is also the area of the circle in which the plane OBC cuts the sphere. Moreover $OM = OP \cos \theta = OA \cos \theta$. Thus we see that *the projection of* OP *on* OA *bears the same ratio to* OA *that the projection on the plane* OBC *of the section conjugate (at right angles) to* OP *bears to the section by* OBC.

The proposition thus proved for the sphere may be extended to the ellipsoid if we remember that the ratio of areas on the same or parallel planes is unaltered by the strain. The projections must now be parallel projections; that is, p is projected on oa by the line pm parallel to the plane obc; and the conjugate area must be projected on obc by lines parallel to oa. The projected area will then bear the same ratio to the section by obc that om does to oa.

We shall use this proposition in representing the strained position or the displacement of any vector, just as we used the corresponding property of the ellipse.

At any point of a sphere, all the straight lines which touch the surface lie in one plane, called the tangent plane at that point. The same thing is therefore true for the ellipsoid.

Now let a be a point on an ellipsoid, such that either oa is the greatest distance from the centre, and the distance of all points immediately surrounding it is less than oa, or else some of these are equal to oa but none greater. There must clearly be such a point on the surface. If now we cut the surface by a series of planes through oa, the tangent lines to all these sections at a will be perpendicular to oa; for each of these sections is either an ellipse or a circle, and in the case of an ellipse oa must be its semi-major axis. Consequently the tangent plane at a is perpendicular to oa. Hence if ob and oc are the axes of the section made by the plane through o perpendicular to oa, the three lines oa, ob, oc form a system of three conjugate diameters at right angles to one another. These are called *axes* of the ellipsoid. The planes containing them two and two are called *principal planes* of the surface, which is evidently symmetrical in regard to each of these planes.

If oa is equal to either ob or oc—say to ob—then the section of the surface by the plane oab is a circle (being an ellipse with equal axes) and any two diameters at right angles in that plane are conjugate diameters. The surface may then be made by rotating an ellipse about its shorter axis oc. It is called an *oblate spheroid*, or *oblatum*. This is approximately the figure of the Earth.

If ob and oc are equal, both being shorter than oa, the section obc is a circle, and any two rectangular diameters in that plane are conjugate. This surface may be made by rotating an ellipse about its *longer* axis oa; it is called a *prolate spheroid*, or *prolatum*. It has two foci (those of the rotating ellipse) the sum of whose distances from any point of the surface is equal to the major axis.

If, on the contrary, oa, ob, oc are all unequal, and in descending order of magnitude, we may derive the ellipsoid from a sphere having the same centre o and radius oa, by reducing all lines parallel to ob in the ratio $OB : ob$, and all lines parallel to oc in the ratio $OC : oc$. It will then be clear that every set of semiconjugate diameters op, oq, or on the same side of the plane oab lies entirely *outside* the solid angle formed by the rectangular lines oP, oQ, oR of which they are the strained positions. Hence the axes are the *only* set of conjugate diameters at right angles to one another.

It follows that in any homogeneous strain of a solid, there are three directions at right angles to one another, which remain perpendicular after the strain; namely those which become the directions of the axes of the ellipsoid into which the strain converts any sphere. Lines drawn through any point in these directions are called *principal axes* of the strain, and the elongations along them are called *principal elongations.*

If the axes remain parallel to their original directions, the strain is called *pure;* if they are turned round, it is accompanied by rotation.

REPRESENTATION OF PURE STRAIN BY ELLIPSOID.

We may now represent (in the case of a pure strain) the strained position of any vector by means of an ellipsoid, in a way entirely analogous to our previous representation of a plane strain by means of an ellipse. Let the principal elongations be e, f, g, and let $p = 1 + e$, $q = 1 + f$, $r = 1 + g$, so that the principal axes of the strain are multiplied by p, q, r respectively. Now construct an ellipsoid with semiaxes a, b, c such that the strained length of a shall represent the area of the section by the plane of b, c, and so for the others; that is, so that $pa = \pi bc$, $ql = \pi ca$, $rc = \pi ab$. This will be effected if we make $a\sqrt{p} = b\sqrt{q} = c\sqrt{r} = \sqrt{(pqr)} : \pi$. Thus the axes of the ellipsoid must be taken inversely proportional to the square roots of p, q, r, which agrees with the rule for the ellipse.

This being so, it follows that the strained position of any vector op represents the area of the section by the conjugate diametral plane; that is to say, it is at right angles to this area, and contains as many linear centimeters as the area contains square ones. For since the projection of that area on the plane obc is to πbc as om to oa, it follows that the strained position of om represents that projection; and similarly the strained positions of mn and np represent the projections on coa, aob. The strained position of op is the vector-sum of these three lines, and therefore represents the area of which they represent the projections.

Thus *the strained position of any radius of this ellipsoid is a vector representing the area of the conjugate section.*

We may easily see that the *volumes* of all portions of the solid are altered in the same ratio by the strain. For we may suppose these volumes cut up into small cubes by systems of planes at right angles, so as to leave pieces over at the boundaries. These cubes will be changed into equal and similar parallelepipeds, and therefore a volume made up of any number of the cubes will be altered in the same ratio as any one cube. Now any volume may be made up of cubes with an approximation which can be made as close as we like by taking the cubes small enough. Hence the proposition follows.

Now the cylinder standing on any diametral section of a sphere, and bounded by the tangent planes parallel to that section, is evidently of constant volume, whatever diametral plane be taken. Hence, in the ellipsoid also, if we draw through every point of a diametral section a line parallel to the conjugate diameter, these lines will constitute a cylinder such that the volume of it enclosed by the two tangent planes parallel to the diametral section is constant, and therefore equal to $2\pi abc$, its value when the section is one of the principal planes. The volume of a cylinder being the product of its base and height, and the height of this one being the perpendicular distance between the parallel tangent planes, that is, twice the perpendicular on either from the centre; it follows that the perpendicular on a tangent plane, multiplied by the area of the parallel diametral section, is equal to a constant, h. Hence if ot be the perpendicular on the tangent plane at p, the strained position of op is along ot and equal in length to $h : ot$.

PROPERTIES OF HYPERBOLOID.

We have hitherto supposed p, q, r to be of the same sign, which, for reasons already mentioned, is the case in all actual strains. If, however, we wish to represent in this way, not the strained position of op, but the displacement of p, we must make the squared axes of our surface inversely proportional to e, f, g, the principal elongations.

So long as these are of the same sign, the displacement may be represented in the way just described; namely, we can construct an ellipsoid so that the displacement of any point p on it shall be a vector representing the area of section conjugate to op. But when one is of a sign different from the other two, we require other surfaces, which shall be now described.

If we make a hyperbola rotate about its transverse axis aa', we obtain a surface of two sheets, each sheet being generated by a branch of the hyperbola. This surface is called a *hyperboloid of revolution of two sheets.*

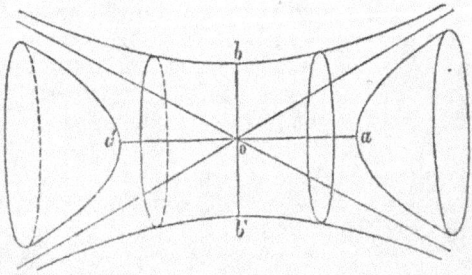

By the same revolution the conjugate hyperbola generates a surface of *one* sheet, the two branches changing places after a rotation through two right angles. This surface is called a *hyperboloid of revolution of one sheet.*

Now let the whole figure be subjected to a uniform strain of any kind; then the surfaces will no longer be surfaces of revolution. They are then called *hyperboloids* of one and two sheets respectively; and in this particular relation are called *conjugate.* To every hyperboloid of one sheet there is a conjugate hyperboloid of two sheets, and *vice versa.* The properties of these more general hyperboloids may be derived from the particular case of the surfaces of revolution, just as those of the ellipsoid are derived from the sphere.

Then, the asymptotes of the revolving hyperbola generate a right cone, called the *asymptotic cone,* towards which the surface approaches indefinitely as it gets further away from the centre. The strain will convert this cone into an *oblique* cone (a cone standing on a circle with the vertex not directly over the centre of the circle) which will still be asymptotic. The shape of this cone determines the shape of the two conjugate surfaces.

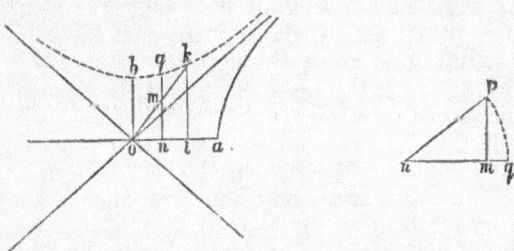

Every central section of a hyperboloid of revolution is a conic; an ellipse when only the one-sheeted surface is cut, a hyperbola when both of the conjugate surfaces are cut. Let the section be made by a plane through *ok* perpendicular to the plane of the paper. When the point q is brought by the rotation to the position p vertically above m, $np^2 = nm^2 + mp^2$, or $mp^2 = nq^2 - nm^2$. Now

$$\frac{nq^2}{ob^2} - 1 = \frac{on^2}{oa^2} = \frac{on^2}{ol^2} \cdot \frac{ol^2}{oa^2} = \frac{on^2}{ol^2}\left(\frac{lk^2}{ob^2} - 1\right) = \frac{nm^2}{ob^2} - \frac{on^2}{ol^2}.$$

Therefore $\dfrac{mp^2}{ob^2} + \dfrac{om^2}{ok^2} = 1$, or the point p lies on an ellipse having *ok* for its semi-major axis, and a line *oc* perpendicular to the plane, of length equal to *ob*, for its semi-minor axis. In a precisely similar way it may be shewn that a central section of the two-sheeted surface is a hyperbola, whose conjugate hyperbola is the section of the conjugate surface by the same plane.

Since after the strain an ellipse remains an ellipse and a hyperbola a hyperbola, it follows that a central section of any hyperboloid is a conic, is an ellipse when only the one-sheeted surface is cut, and a hyperbola when both the conjugate surfaces are cut.

Now take any point p on a hyperboloid of two sheets, and draw through the centre a plane oqc parallel to the tangent plane at p; this will cut the conjugate hyperboloid in an ellipse. Let the whole figure receive a shear, by sliding over one another the planes parallel to the tangent plane at p, until op becomes perpendicular to them; then elongate all lines parallel to the shorter axis oc of the ellipse until the ellipse becomes a circle. Then every section by a plane through op will be a hyperbola of which op is the transverse semi-axis, because it is perpendicular to the tangent at p. Consequently the other axis is in the plane of the circle and equal to its diameter; that is to say, all these hyperbolas have the same axes, and are therefore equal and similar. Hence the conjugate surfaces have been converted by this strain into surfaces of revolution.

In this state of the figure it is clear (1) that the tangent planes at points on the section by obc are parallel to op; (2) that all sections parallel to obc are circles having their centres on op. Hence in general if p be any point on a hyperboloid of two sheets, and oqc a diametral plane parallel to the tangent plane at p, the tangent planes at all points of the section of the conjugate surface by oqc are parallel to op, and all sections parallel to oqr are similar and similarly situated ellipses having their centres on op. If we draw through o a plane opr parallel to the

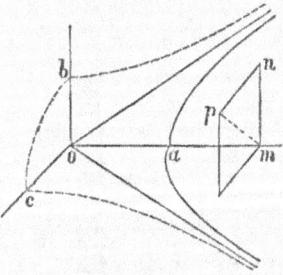

tangent plane at any point q of the section oqr, this will cut the hyperboloid of two sheets in a hyperbola, the tangent plane at every point of which will be parallel to oq, since any such point might be taken for the point p. Hence if we take any two conjugate diameters oq, or of the section oqr, the three lines op, oq, or are such that the tangent plane at the extremity of each is parallel to the other two. These lines are called a set of *conjugate diameters* of either of the two surfaces; one of them always meets the hyperboloid of two sheets, and the other two meet the hyperboloid of one sheet.

Now in the surface of revolution, any section through oa being a hyperbola whose semi-axes are equal to oa and ob, we have $\dfrac{om^2}{oa^2} - \dfrac{pm^2}{ob^2} = 1$; or since $pm^2 = mn^2 + np^2$, we have $\dfrac{om^2}{oa^2} - \dfrac{mn^2}{ob^2} - \dfrac{np^2}{oc^2} = 1$. Since each of these ratios is unaltered by a homogeneous strain, the equation is equally true for any hyperboloid of two sheets, if now oa, ob, oc form a set of conjugate semi-diameters in the sense just explained. It may be shewn in the same way that in a hyperboloid of one sheet we should find

$$- \frac{om^2}{oa^2} + \frac{mn^2}{ob^2} + \frac{np^2}{oc^2} = 1.$$

The two equations may also be written respectively

$$\frac{x^2}{a^2} - \frac{y^2}{b^2} - \frac{z^2}{c^2} = \pm 1,$$

x, y, z being written for om, mn, np, as before.

It follows immediately that any plane parallel to aob cuts the surfaces in two hyperbolæ whose common centre is on oc, the asymptotes of all being parallel.

DISPLACEMENT-QUADRIC.

It shall now be proved that any homogeneous strain of a solid may be represented by means of a central quadric surface, namely, either an ellipsoid or a pair of conjugate hyperboloids, in the following manner. The displacement of any point p of the surface, relative to its centre o, will be at right angles to the diametral section conjugate to op, and will contain as many centimeters of length as that section contains square centimeters of area. For this purpose it is necessary to shew that the hyperboloids have the same property which we proved true for the ellipsoid; namely that if a section A and its conjugate diameter α be respectively projected upon a section B and its conjugate diameter β, by lines parallel to β and B respectively, the ratio of the projection of A to B is equal to the ratio of the projection of α to β. We shall prove this first for surfaces of revolution, and then extend it to the other surfaces by a homogeneous strain.

When the central section is a hyperbola, we cannot properly speak of its area at all. In this case we shall suppose it to be replaced by an ellipse having the same axes; so that in general, if the semi-axes of an ellipse or hyperbola are a, b, the area is always to be reckoned as πab.

Let the figure represent two conjugate hyperbolæ, which, by revolving about the axis aa', are to generate a pair of conjugate hyperboloids of revolution. The diame-
tral section conjugate to op is made by a plane through qq' perpendicular to the paper. The semi-axes of this section are oq and a line oc perpendicular to the paper equal in length to ob. The projection of the section on ob is therefore $\pi \cdot nq \cdot oc$, and its ratio to the area of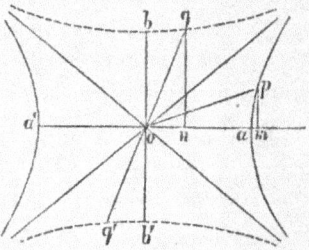
the section obc is $\pi \cdot nq \cdot oc : \pi \cdot ob \cdot oc = nq : ob$. But $nq : ob = om : oa$; thus the projection of section oqc on obc bears the same ratio to section obc that projection of op on oa bears to oa.

In the same way, the section conjugate to oq is made by a plane through op perpendicular to the paper, and its "area" is to be reckoned as $\pi \cdot op \cdot oc$. It follows at once that its projection on the plane oac, namely $\pi \cdot om \cdot oc$, bears the same ratio to section oac' that nq bears to ob.

Passing now to the case of hyperboloids not of revolution, we have proved that any pair of conjugate surfaces may be altered by homogeneous strain into surfaces of revolution, so that *any* given diameter aa' of the hyperboloid of two sheets becomes the axis of revolution. And since the ratios of parallel lengths and of parallel areas are unaltered by the strain, it follows that the property just proved for surfaces of revolution is true for all hyperboloids.

This being so, let e, f, g be principal elongations of a homogeneous strain, and let a, b, c be three lengths such that $a\sqrt{e} = b\sqrt{f} = c\sqrt{g} = \sqrt{(efg)} : \pi$. If any of the quantities e, f, g be negative, we must in this formula consider it replaced by its absolute value. If e, f, g are all of the same sign, construct an ellipsoid with semi-axes a, b, c; but if one is of a sign different from that of the other two, construct a pair of conjugate hyperboloids with the same semi-axes, so that the axes whose elongations are of the *same* sign shall meet the one-sheeted surface, and the remaining axis the two-sheeted surface. The relation between x, y, z in the quadric surface or surfaces thus constructed, whether ellipsoid or hyperboloids, is

$$\pi (ex^2 + fy^2 + gz^2) = \pm efg,$$

as may be seen by comparing the values just given for a, b, c with the equation $\frac{x^2}{a^2} \pm \frac{y^2}{b^2} \pm \frac{z^2}{c^2} = \pm 1$. The surface for which $\pi (ex^2 + fy^2 + gz^2) = efg$ is called the *displacement-quadric*. If e, f, g are all positive or all negative, the displacement-quadric is an ellipse; if two of them are positive and one negative, or if one is positive and two negative, it is a hyperboloid of two sheets; but in the latter case we must call in the assistance of the conjugate surface in order to represent the strain.

If then oa, ob, oc be semi-axes of the displacement-quadric, the displacement of the point a is ea which is πbc, the area of the conjugate section; and this displacement is along oa, and therefore normal to the area of that section. When the displacement-quadric is a hyperboloid, elliptic and hyperbolic areas must be regarded as having different signs; but *which* sign is to be attributed to each depends on the signs of e, f, g, and it will be found in fact that the elliptic area is always of the same sign as the product efg.

Since the displacements of a, b, c are vectors representing the conjugate areas, it follows that the displacement of any point! p on the displacement-quadric or its conjugate surface is a vector representing the area of section conjugate to op. For we have shewn that the *components* of that area, namely its projections on the principal planes, bear the same ratio to the principal areas πbc, πca, πab, that the components of op, namely its projections on the axes, bear to those axes. Now if om be the projection of op on oa, the displacement of m is $\dfrac{om}{oa}$ × the displacement of a, that is, it is $\dfrac{om}{oa} \times \pi bc$. Consequently the displacement of m is a vector representing the projection on obc of the area conjugate to op. Now the displacement of p is the resultant of the displacements of its projections on the axes; and therefore it represents the area which is the resultant of the three projections here considered, namely, the area of section conjugate to op.

The case of a point k lying on the *asymptotic cone* of the displacement-quadric requires some explanation. In that case the length of the line drawn in the direction ok to meet the surface is infinite, and the displacement of its end is infinite also. The conjugate section is made by a plane through ok touching the asymptotic cone, which cuts the conjugate surface in two parallel straight lines. In the case of sur-

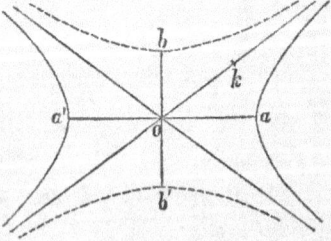

faces of revolution it is clear that the distance between these lines is bb'; they lie on either side of ok in a plane through it perpendicular to the paper. Thus the displacement of p (the infinitely distant point on ok) is $\pi . ob . op$, perpendicular to ok in the plane of the paper. Hence the displacement of k is $\pi . ob . ok$ in the same direction. And generally the displacement is $\pi . ok$ multiplied by half the breadth of the conjugate section.

In any other case if ot be the perpendicular on the tangent plane at p, the displacement of p is parallel to ot and equal to $\pi abc : ot$. For the perpendicular on a tangent plane, multiplied by the area of the parallel diametral section, is constant, and therefore equal to πabc. This follows at once for surfaces of revolution from the corresponding property of the hyperbola; and it is extended to any hyperboloids by the consideration that all volumes are altered in the same ratio by a homogeneous strain. We shall write H for πabc or $efg : \pi^2$, so that displacement of $p = H : ot$.

LINEAR FUNCTION OF A VECTOR.

Just as in the case of a plane strain, the strained position of a vector or the displacement of its end is said to be a *linear function* of the original vector when the strain is homogeneous. If the displacement of the end of ρ be denoted by $\phi(\rho)$, the strained position of it is $\rho + \phi(\rho) = (1 + \phi)\rho$. When the strain is pure, ϕ is said to be a *pure function*.

Let i, j, k be three unit-vectors at right angles to one another, and let

$$\phi i = ai + hj + g'k,$$

$$\phi j = h'i + bj + fk,$$

$$\phi k = gi + f'j + ck.$$

Then if $\rho = xi + yj + zk$, we shall have $\phi\rho = x\phi i + y\phi j + z\phi k$, so that the function of every vector can be expressed in terms of these, and the strain is entirely specified by means of the nine quantities $a, b, c, f, g, h, f', g', h'$. The equations just written down are sometimes conveniently abbreviated as follows :—

$$\phi i, \phi j, \phi k = \begin{pmatrix} a & h & g' \\ h' & b & f \\ g & f' & c \end{pmatrix}(i,\, j,\, k), \text{ or } \phi = \begin{pmatrix} a & h & g' \\ h' & b & f \\ g & f' & c \end{pmatrix}$$

and the form $\begin{pmatrix} a & h & g' \\ h' & b & f \\ g & f' & c \end{pmatrix}$ is also called a *matrix*. Thus

every strain has a certain matrix belonging to it, which serves to define the strain by means of its displacement function.

When the strain is pure, the *scalar product* of *op* and the displacement of p is $- H$, if p is a point on the displacement-quadric. If *pq* is the displacement of p, the scalar product is

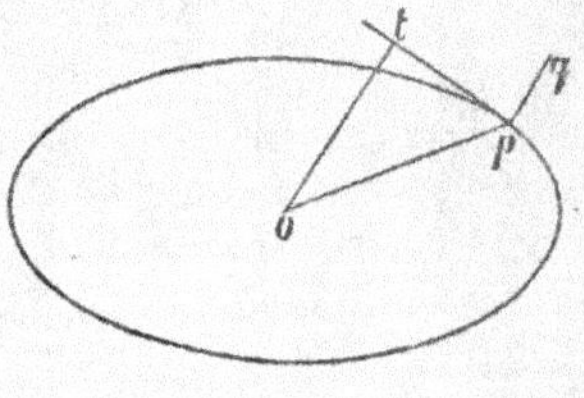

$$op \cdot pq \cos opq;$$

but we know that

$$pq = H : ot,$$

and

$$op \cos opq = - op \cos pot = - ot,$$

which proves the theorem. Hence if ρ is the step from the centre to a point on a quadric surface, $S\rho\phi\rho = -H$, whence H is π times the product of the semi-axes of the surface.

The scalar product of two vectors is the negative sum of the products of their components along the axes. For

$op \cos poq$ is the projection of
op on oq, which is the sum of
the projections of on, nm, mp
on oq. Let x, y, z be the
components of op, x_1, y_1, z_1
those of oq, and r, r_1 the
lengths of op and oq. Then
if oq makes angles α, β, γ with
oX, oY, oZ, we must have

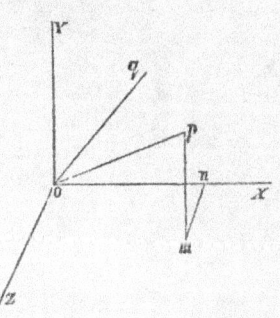

$$\cos \alpha = x_1 : r_1, \quad \cos \beta = y_1 : r_1,$$
$$\cos \gamma = z_1 : r_1.$$

And $op \cos poq = on \cos qoX + mp \cos qoY + nm \cos qoZ$

$$= x \cos \alpha + y \cos \beta + z \cos \gamma = xx_1 + yy_1 + zz_1 : r_1.$$

Therefore $Sop . oq = - op . oq \cos poq = - (xx_1 + yy_1 + zz_1).$

Let ρ, σ be two vectors from the centre to points on
the displacement-quadric; then $S\rho\phi\sigma = S\sigma\phi\rho$. Since
$\phi\sigma$ is a vector at right angles to the
section conjugate to σ, whose length
represents the area of that section,
$S\rho\phi\sigma$ will be the volume of a cylinder
standing on that section and having ρ
for its axis. We have to shew that this
volume is equal to that of a cylinder
standing on the section $\phi\rho$ and having σ for its axis.
This proposition is obvious when the quadric is a sphere;

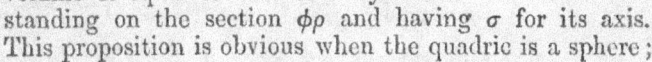

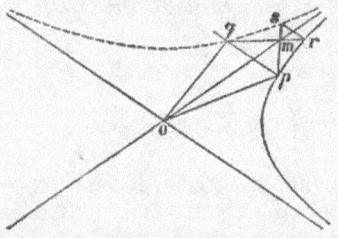

whence it follows for the ellipsoid by means of a homogeneous strain. For a hyperboloid of revolution we may

deduce it from the property already proved for the hyperbola, that the triangles ops, orq are equal if op is conjugate to oq and or to os. For the section conjugate to op is $\pi . oc . oq$ if oc is the semi-diameter perpendicular to the plane of the paper; and the volume of a cylinder standing on this with or for axis will be $\pi . oc . \times$ twice triangle orq, which is equal to $\pi . oc \times$ twice triangle ops, the volume of a cylinder standing on the section ocs with op for axis. From this the proposition follows by a homogeneous strain for all hyperboloids.

Analysing this proposition to the expressions just given for ϕi, ϕj, ϕk, we may shew that *when the strain is pure $f = f'$, $g = g'$, $h = h'$*. For we have

$$\phi i = ai + hj + g'k, \quad \phi j = h'i + bj + fk;$$

and therefore

$$Si\phi j = -h', \quad Sj\phi i = -h.$$

Thus a pure strain depends only on the six quantities a, b, c, f, g, h, whereas a strain not known to be pure is specified by nine quantities.

We may now prove that

$$S\rho\phi\rho = S(xi + yj + zk)(x\phi i + y\phi j + z\phi k)$$

$$= -(ax^2 + by^2 + cz^2 + 2fyz + 2gzx + 2hxy),$$

so that

$$ax^2 + by^2 + cz^2 + 2fyz + 2gzx + 2hxy = H.$$

This is the relation which holds good between x, y, z for all points on the displacement-quadric.

VARYING STRAIN.

In a homogeneous strain, if we suppose one point of the body to be at rest, and draw any straight line through it, the displacements of points on this line will be all in the same direction and proportional to the distance from the fixed point. Hence *the relative displacement of the two points at a unit distance along this line will be equal to the rate of change of the displacement per unit distance as we go along the line.* Let σ represent the displacement of any point, and let α be the step from a point p to a point q. The symbol $\partial_\alpha \sigma$ shall mean the change in σ due to the step α; it will therefore be the displacement of q relative to p. This is what we have previously denoted by $\phi \alpha$; so that we may now write $\partial_\alpha \sigma = \phi \alpha$, where σ is the displacement and ϕ the displacement-function.

It thus appears that the strain at any point of a body does not depend on the actual displacement, but on the variation of the displacement in the neighbourhood of the point. When a body is subject to strain which is not homogeneous, we can find, for every point of the body, a homogeneous strain such that the rate of change of the displacement in every direction due to the homogeneous strain shall be the same as the rate of change of displacement in that direction in the actual condition of the body. This homogeneous strain is called the strain *at* the point. It varies, in general, from one point to another.

Consider now a point p of the body, and draw a line pq through it, which shall be called α. As we start from p to go along pq, there is a certain rate of change of the displacement σ. Let $\partial_\alpha \sigma$ represent what the difference in displacement of q and p would be *if this rate of change were uniform from p to q.* That is to say, $\partial_\alpha \sigma$ is the displacement of q relative to p in the homogeneous strain which coincides with the actual strain at p. Hence if ϕ is the displacement-function of this strain, $\partial_\alpha \sigma = \phi \alpha$; where now $\partial_\alpha \sigma$ means the rate of change as we go in the direction α, multiplied by the length of α. For example, $\partial_i \sigma$ is the same thing as we have called $\partial_x \sigma$; so $\partial_j \sigma = \partial_y \sigma$, and $\partial_k \sigma = \partial_z \sigma$, because i, j, k are of unit length.

Hence if the components of σ are u, v, w, so that

$$\sigma = ui + vj + wk,$$

we shall find

$$\phi i = \partial_x \sigma = \partial_x u . i + \partial_x v . j + \partial_x w . k,$$
$$\phi j = \partial_y \sigma = \partial_y u . i + \partial_y v . j + \partial_y w . k,$$
$$\phi k = \partial_z \sigma = \partial_z u . i + \partial_z v . j + \partial_z w . k,$$

and consequently the matrix of the function ϕ is

$$\begin{pmatrix} \partial_x u, & \partial_x v, & \partial_x w \\ \partial_y u, & \partial_y v, & \partial_y w \\ \partial_z u, & \partial_z v, & \partial_z w \end{pmatrix}$$

CHAPTER II.   STRAIN-VELOCITIES.

HOMOGENEOUS STRAIN-FLUX.

WE have already investigated all those velocity-systems which are consistent with rigidity, and shewn how to compound them together. It is probable, however, that no body in nature is ever rigid for so much as a second together. The most solid masonry is constantly transmitting vibrations which it receives from the earth's surface and from the air; these vibrations constitute minute changes of shape. Other minute changes of shape are due to the varying position of attracting bodies, such as the moon. The spins and twists, therefore, which we have investigated are to be regarded as *ideal* motions, to which certain natural motions more or less closely approximate. The motions of fluids, however, such as water or air, are not even approximately consistent with rigidity, and to describe these we must consider some other velocity-systems. As before, we have to describe ideal motions, which can be dealt with by exact methods, but which only approximately represent the motions which actually take place.

Imagine an elastic string, one end of which is fixed, while the other end moves uniformly along a straight line passing through the fixed end, so that the string is always stretched along the same line. If the strain is always homogeneous, the velocities of any two points on the string will at any moment be proportional to their distances from the fixed end.

Now consider an infinite plane surface, with air on one side of it; and let those particles of air which lie along any straight line perpendicular to the plane be moving like the particles of the elastic string just considered; that is to say, let the velocity at every point be perpendicular to the plane and proportional to the distance from it. Let a similar motion take place on the other side of the plane, but in the opposite direction; that is, so that both motions are towards the plane, or both away from it. If the velocity at distance x from the plane is ex, and if we suppose the velocity of every particle to remain uniform for one second, then at the end of that second there will be produced a uniform elongation e perpendicular to the plane. For the moment, we may call this velocity-system a *stretch* perpendicular to the given plane.

Take now three planes intersecting at right angles in a point o, and combine together at every point of space the velocities due to stretches e, f, g perpendicular to these three planes respectively. We shall then have a velocity-system such that if the velocity of every particle remains uniform for one second, there will be produced a pure homogeneous strain of which e, f, g are the principal elongations.

Lastly, combine with this velocity-system a *spin* about some axis passing through the point o. The resultant velocity-system has then the following properties.

1. The point o is at rest.

2. The velocities of all points lying in a straight line through o are parallel, and proportional to the distance from o.

3. If the velocity of every point be kept uniform for one second, there will be produced at the end of that second a homogeneous strain.

Let σ be the displacement at the end of one second of a particle whose position-vector from o as origin was originally α, then $\sigma = \phi\alpha$, where ϕ is the displacement-function of the homogeneous strain. Hence the equation to the uniform motion of this particle during the second is

$$\rho = \alpha + t\phi\alpha.$$

Consequently at the time $t = 0$, the particle whose position-vector is α has the velocity $\phi\alpha$. The motion is therefore such that *the velocity at every point is a linear function of the position-vector of the point.* Such a velocity-system may be called a *homogeneous strain-flux.* We may formally define it as follows;

If at any instant the velocity-system of a body be such that by keeping the velocity of each point uniform for one second we should produce a homogeneous strain ϕ, then at that instant the body is said to have the homogeneous strain-flux ϕ.

If we combine with this velocity-system a translation equal and opposite to the velocity of any point p, the resultant will be a new homogeneous strain-flux with the point p for centre. For if we keep all velocities constant for a second, we shall produce a homogeneous strain together with a translation restoring p to its place; that is, a homogeneous strain in which p is not moved.

It is clear that the resultant of two homogeneous strain-fluxes is again a homogeneous strain-flux; but in this term we must include as special cases the motions consistent with rigidity. A *twist* may be regarded as a homogeneous strain-flux whose centre o is infinitely distant; in the still more special case of a *spin*, the centre is indeterminate, being any point whatever on the axis.

The latter case is distinguished by the function ϕ being a *skew* function. For let the spin $\omega = pi + qj + rk$, then the velocity of any point whose position-vector is ρ will be $V\omega\rho$. Consequently we have $\phi\rho = V\omega\rho$, and therefore

$$\phi i = V\omega i = \quad\quad + rj - qk,$$
$$\phi j = V\omega j = - ri \quad\quad + pk,$$
$$\phi k = V\omega k = + qi - pj \quad\quad ;$$

so that the matrix of ϕ is

$$\begin{pmatrix} 0, & +r, & -q \\ -r, & 0, & +p \\ +q, & -p, & 0 \end{pmatrix}.$$

We may now separate any given homogeneous strain-flux into the pure part of it and the spin. For it is evident that

$$
\begin{pmatrix} a, & h, & g' \\ h', & b, & f \\ g, & f', & c \end{pmatrix} = \begin{pmatrix} a, & \tfrac{1}{2}(h+h'), & \tfrac{1}{2}(g+g') \\ \tfrac{1}{2}(h+h'), & b, & \tfrac{1}{2}(f+f') \\ \tfrac{1}{2}(g+g'), & \tfrac{1}{2}(f+f'), & c \end{pmatrix}
$$

$$
+ \begin{pmatrix} 0, & \tfrac{1}{2}(h-h'), & \tfrac{1}{2}(g'-g) \\ \tfrac{1}{2}(h'-h), & 0, & \tfrac{1}{2}(f-f') \\ \tfrac{1}{2}(g-g'), & \tfrac{1}{2}(f'-f), & 0 \end{pmatrix}.
$$

Here the first of the matrices on the right hand belongs to a pure function, and the second to a spin, whose components are $\tfrac{1}{2}(f-f')$, $\tfrac{1}{2}(g-g')$, $\tfrac{1}{2}(h-h')$. The resolution cannot be effected in any other way; for to change the spin into any other (not about a parallel axis) we must combine a spin with it. The resultant of the pure strain-flux and of this spin reversed will be no longer a pure strain-flux.

CIRCULATION.

Consider a plane curve joining two points p and q. Let a line be drawn through every point of the curve, perpendicular to its plane, representing the component of velocity along the tangent to the curve at that point. All these lines will trace out a strip or riband standing on the curve. The area of this strip is called the *circulation* along the curve from p to q. When the resolved part of the 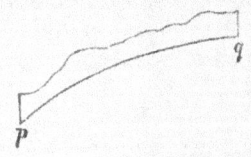 velocity is in the direction from q to p, it is to be drawn *below* the plane, and that part of the area is to be reckoned negative. Hence the circulations from p to q and from q to p are equal in magnitude but of opposite sign.

The circulation may also be described as follows. Divide the length of the curve into small pieces, of which $\delta\lambda$ is one. Let σ be the velocity of some point included in the piece $\delta\lambda$, then $-S\sigma\delta\lambda$ will be the resolved part of this velocity along the curve, multipled by the length of $\delta\lambda$. The sum $-\Sigma S\sigma\delta\lambda$ of such quantities for all pieces of the curve may be made to approximate as near as we please to a quantity $-\int S\sigma d\lambda$ by increasing the number and diminishing the length of the pieces. This quantity $-\int S\sigma d\lambda$ is called the circulation along the curve from p to q. The second definition is equally applicable to a non-plane curve, on which we cannot draw a riband which shall represent the circulation by its area.

If we suppose the point q to move along the curve pq with unit velocity, the rate of change of the circulation from p to q will be the component along the tangent at q of the instantaneous velocity of the body at q. For if this component remained constant over a unit length of the curve, the change of circulation would be the component multiplied by the unit of length. Thus if s denote the length of the arc pq, and C the circulation from p to q, $\partial_s C = v\cos\theta$, where $v =$ velocity at q, and $\theta =$ angle it makes with the tangent to pq.

In general, if σ be any vector which has a definite value at every point of space, the quantity $-\int S\sigma d\lambda$ is called the *line-integral* of σ along the curve λ; so that we may say that the circulation is the line-integral of the velocity.

If an area be divided into parts, the circulation round the whole area is equal to the sum of the circulations round the parts. The area $abcd$, for example, is made up of abc and acd. The circulation round abc is made up of that along abc and that along ca. The circulation round acd is made up of that along ac and that along cda. Now the circulation along 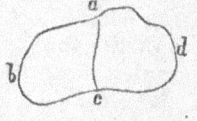 ca is equal and opposite to that along ac, so that when we put the circulations round the parts together, these two portions destroy one another, and the sum is the circulation round $abcd$. The same reasoning applies to any number of parts. It it clear that the proposition holds equally good, whether the areas are on a plane or on any other surface.

We may conveniently write (ab) for the circulation along ab. Thus $(abcd) = (abca) + (acda)$.

In a homogeneous strain-flux, the circulation round two closed curves is the same if one can be made to coincide with the other by a step of translation. For if the positions of two corresponding points differ by the constant vector a, then the velocities differ by the constant quantity ϕa; and the difference of the circulations is merely the length of ϕa multiplied by the projection of the closed curve on a line parallel to ϕa, which is of course zero.

Since the circulation round a closed curve is thus unaltered by the same velocity being given to all its points, we may if we like reduce any one point to rest, without altering the circulation round any closed curve.

The circulation round any two parallelograms of the same area is the same. We may change $abdc$ into $abfe$ by adding ace and subtracting bdf; and the circulation round these two triangles is the same. By repeating this process we may make one parallelogram into a translation of any other of equal area. By *equal area* is of course implied that they are in the same or parallel planes.

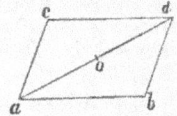

The circulation round any parallelogram is double of that round a triangle of half its area. Let o, the middle point of ad, be brought to rest. Then the circulation along ad is zero, and the velocities at corresponding points of ab and dc being equal and opposite, the circulation along ab is equal to that along dc; similarly that along bd is equal to that along ca. Thus $(ab) + (bd) + (da) = (ad) + (dc) + (ca)$, or the circulations round the triangles abd, adc are equal, and therefore each half of the circulation round $abdc$.

It follows that the circulation round any two triangles of the same area is the same.

Hence *the circulations round any two areas in the same or parallel planes are proportional to those areas.*

For we may replace each of them by a polygon with short rectilineal sides, and these polygons may then be divided into small equal triangles. The areas will be nearly as the numbers of these triangles with an approximation which can be made as close as we like by making the triangles small enough. But the circulation round each polygon is the sum of the circulations round its component triangles; therefore the two circulations are also as the numbers of the triangles approximately, and therefore as the two areas exactly.

If α, β are the sides of a parallelogram, the circulation round it is $S\beta\phi\alpha - S\alpha\phi\beta$. Let $\alpha = ab$, $\beta = ac$. Then the sum of the circulations along ab and dc is the difference of those along ab and cd; which is the length of α multiplied by the resolved part of $\phi\beta$ along it, or $-S\alpha\phi\beta$. Similarly the sum of the circulations along bd and ca is the

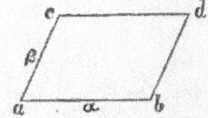

difference of those along bd and ac, which is seen in the same way to be $S\beta\phi\alpha$. Hence the proposition.

The circulation round any plane area is equal to twice the product of the area by the component of spin perpendicular to it. A unit area in the plane oXY is the square whose sides are i, j. Now $\phi i = ai + hj + g'k$, $\phi j = h'i + bj + fk$; therefore $Si\phi j - Sj\phi i = h - h'$, which is twice the component of spin round oZ. Now any plane whatever may be taken for the plane of oXY; whence the proposition.

STRAIN-FLUX NOT HOMOGENEOUS.

In the case of a homogeneous strain-flux, if we take any point p of the body and draw a straight line pq through it, the velocities of points on this line, relative to p, will all be parallel and proportional to the distance from p along the line. Consequently the rate of change of the velocity, as we go along the line pq, is constant.

When the strain-flux is not homogeneous, this rate of change of the velocity will no longer in general be constant. But we may imagine a homogeneous strain-flux

which is such that the rate of change of velocity due to
it. in any direction, is the same as the rate of change at p
when we are moving in that direction in the actual condition of the body. This homogeneous strain-flux will
then be called the strain-flux at p. It will in general vary
from one point of the body to another.

In order that there may be a strain-flux at p at all, it
is necessary that the velocity should change gradually as
we pass through p in any direction. That is to say, there
must be a rate of change up to p, and a rate of change on
from p, and these must be equal. When this is the case,
the entire strain-flux of the body may be said to be *elementally homogeneous*, or *homogeneous in its smallest parts*.
Any small portion of the body moves with an approximately homogeneous strain-flux, and the approximation
may be made as close as we like by taking the portion
small enough. But if one portion of the body is sliding
over another portion with finite velocity, this is not the
case. In crossing the common surface of the two portions,
we should find a sudden jump in the velocity. Such dis-
continuities have to be separately considered.

Let now α be any vector drawn through the point p;
and let $\partial_a \sigma$, as before, mean the change that would be produced in σ by passing from one end of α to the other, if
the rate of change per unit length remained uniformly
what it actually is at p. Then the strain-flux at p has a
velocity-function ϕ such that $\partial_a \sigma = \phi \alpha$. If therefore

$$\sigma = ui + vj + wk,$$

the matrix of ϕ is

$$\begin{pmatrix} \partial_x u & \partial_x v & \partial_x w \\ \partial_y u & \partial_y v & \partial_y w \\ \partial_z u & \partial_z v & \partial_z w \end{pmatrix}.$$

Consequently the spin ω is

$$\tfrac{1}{2}\{(\partial_y w - \partial_z v)i + (\partial_z u - \partial_x w)j + (\partial_x v - \partial_y u)k\}.$$

It follows from this formula that if two velocity-systems
are compounded together, the spin at any point in the
resultant motion is the resultant of the two spins in the
component motions.

LINES OF FLOW AND VORTEX-LINES.

At every instant a moving body (to fix the ideas, consider a mass of water) has a certain velocity-system, i.e. every point in the body has a certain velocity σ. A curve such that its tangent at every point is in the direction of the velocity of that point is called a *line of flow*. It is clear that a line of flow can be drawn through any point of the body, so that at every instant there is a system of lines of flow. If the body has a motion of translation, the lines of flow are straight lines in the direction of the translation. If it rotates about an axis, the lines of flow are circles round the axis. If fluid diverges in all directions from a point, the lines of flow are straight lines through that point.

It is important to distinguish a line of flow from the actual path of a particle of the body. A line of flow relates to the state of motion at a given instant, and in general the system of lines of flow changes as the motion goes on. Thus while the path of a particle touches at every instant the instantaneous line of flow which passes through the particle, it does not in general coincide with any line of flow. The particular case in which the system of lines of flow does not alter, and in which, therefore, each of them is actually the path of a stream of particles, is called *steady motion*. In that case, the lines of flow are called stream-lines.

Thus, if a rigid body move about a fixed point, we know that its velocity-system at every instant is that of a spin about some axis through the fixed point, and consequently the lines of flow are circles about that axis. But in general the axis changes as the motion goes on, and the path of a particle of the body is not any of these circles.

If we take a small closed curve, and draw lines of flow through all points on it, the tubular surface traced out by these lines is called a *tube of flow*. In the case of steady motion all tubes of flow are permanent, and the portion of the body which is inside such a tube does not come out of it.

In general, a body has also at every instant a certain *spin-system;* i.e. at every point of the body there is a certain spin ω. In fact, if the strain-flux is elementally homogeneous, there is at every point a homogeneous strain-flux which is the resultant of a pure strain-flux and a spin ω.

A curve such that its tangent at every point is in the direction of the spin at that point is called a *vortex-line.* If we draw vortex-lines through all the points of a small closed curve, we shall form a tubular surface which may be called a *tube of spin;* the part of the body inside the tube is called a *vortex-filament.* In the cases of fluid motion which occur most often in practice, there is a finite number of vortex-filaments in different parts of the fluid, but the remaining parts have no spin.

CIRCULATION IN NON-HOMOGENEOUS STRAIN-FLUX.

If we consider any small area δz, which may be taken to be approximately plane, the strain-flux in its neighbourhood is approximately homogeneous; and if ω be the spin at a point inside of the area, the circulation round the area will be approximately equal to its magnitude multiplied by twice the component of spin perpendicular to it; that is, it will be approximately $- 2S\omega\delta z$, where δz is regarded as a vector representing the area, and therefore perpendicular to it. This approximation is closer, the smaller the area is taken.

Now let *abcd* be any closed contour, whether plane or not, and let us suppose it to be covered by a cap, as *aec,* so that the contour is the boundary of a certain area on the surface of this cap. If this area be divided into a great number of very small pieces, as *f,* each of these may be taken to be approximately plane. And the circulation round 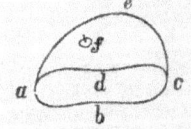 *abcd* will be the sum of the circulations round all the small pieces. Thus it will be approximately equal to $- 2\Sigma\, S\omega\delta z,$

where $\delta\mathbf{z}$ is one of the small pieces, and ω the spin at some point within it. The approximation may be made as close as we like by taking the pieces small enough, and therefore the circulation is exactly equal to the integral $-2\int S\omega\delta\mathbf{z}$. If $\delta\lambda$ be a small piece of the contour $abcd$, we know that the circulation is also equal to $-\int S\sigma d\lambda$; and consequently we have $2\int S\omega d\mathbf{z} = \int S\sigma d\lambda$.

In general, if ω is any vector having a definite value at every point of space, the integral $-\int S\omega d\mathbf{z}$, taken over any area, plane or curved, is called the *surface-integral* of the vector over that area. We may therefore state our proposition thus: the line-integral of the velocity round any contour is equal to twice the surface-integral of the spin over any cap covering the contour.

Let us now draw another cap, agc, covering the contour. Then the surface-integral of the spin over agc must be equal to that over aec, because each is half the line-integral of velocity round $abcd$. But in one case the vectors representing small pieces of area will all be drawn *inwards*, and in the other *outwards*. If then we suppose them all to be drawn outwards, the surface-integral over the entire closed surface $aecg$ will be zero. It is in fact obvious that if we divide the area of any closed surface into small pieces, and suppose each of these to be gone round in a counter-clockwise direction, as viewed from outside, the sum of all their circulations will be zero, since each boundary line is traversed twice, in opposite directions.

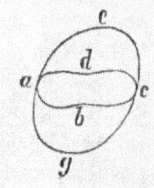

We learn, therefore, that *the surface-integral of the spin over any closed surface is zero*. The closed surface may be that of a body having no holes through it, as in the figure, or it may be that of a body with any number of holes through it; for example, the surface of an anchor-ring, or of a solid figure-of-eight.

Let us now apply this proposition to a portion of a *tube of spin*, cut off at a and b by surfaces of any form.

This closed surface consists of the two ends
at a and b, and of the tubular portion
between them. At every point of the
tubular portion the axis of spin is tangent
to the vortex line through that point,
which lies entirely in the surface; consequently it has no component normal to the
surface. Therefore the tubular portion of
the surface contributes nothing to the
surface-integral. It follows that the sum of the surface-
integrals over the two ends is zero. Now the surface-
integral over either end is half the circulation round its
boundary; but since the lines representing pieces of area
are to be drawn *outwards* in both cases, these boundaries
must be gone round in opposite directions. Since then,
when they are traversed in opposite directions the circulations are equal and opposite in sign, it follows that when
they are traversed in the same direction the circulations
are the same. Or, *the circulation is the same round any
two sections of a tube of spin.*

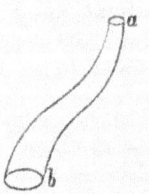

When the tube is small, *the spin at any part of it is
inversely proportional to the area of normal section.* For
then the surface integral over the section is approximately
equal to the spin at any point of it multiplied by the area
of the section; and we have seen that this surface-integral
is constant. Hence a vortex-filament rotates faster in proportion as it gets thinner.

This shews us also that a vortex-filament cannot come
to an end within the fluid, but must either return into
itself, each vortex-line forming a closed curve, as in the
case of a smoke-ring, or else end at the surface of the
fluid, where the velocity no longer changes continuously;
and consequently our previous reasoning does not apply.
Such a vortex-filament may be formed by drawing the
bowl of a teaspoon, half immersed, across the surface of a
cup of tea; the filament goes round the edge of the sub-
merged half of the bowl, and the two ends of it may be
seen rotating as eddies on the surface.

IRROTATIONAL MOTION.

If it is possible to cover a contour by a cap such that there is no spin at any point of it, the circulation round the contour will be zero, since it is equal to twice the surface-integral of the spin, taken over the cap. Let p and q be two points on such a contour $paqb$, then the circulation from p to q is the same along paq as along pbq. For

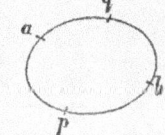

$$(paq) + (qbp) = 0 , \text{ or } (paq) = (pbq).$$

Therefore

Of two paths going from p to q, if it is possible to move one into coincidence with the other without crossing any vortex-line, the circulation along them is the same.

Where there is no spin, the motion is called *irrotational*. If there is no spin anywhere, so that the motion is irrotational throughout all space, the circulation from one point to another is independent of the path along which it is reckoned. Let a point o be taken arbitrarily, then for every point p in the body there is a certain definite quantity, namely, the circulation along any path from o to p. This is called the *velocity-potential* at p. If p be moved about so as to keep its velocity-potential constant, it will trace out a surface which is called an *equipotential* surface. It is clear that we may draw an equipotential surface through every point of space, and in this way we shall have a *system* of equipotential surfaces. There is no circulation along any line drawn on an equipotential surface ; because the circulation from one point to another is equal to the difference of their velocity-potentials. (Circulation from p to $q =$ circ. from o to $q -$ circ. from o to p.)

Suppose, for example, that a body has a motion of translation. Then a plane perpendicular to the direction of motion will be an equipotential surface ; for there is no component of velocity along any line in such a plane, and therefore the circulation along that line is zero. If we choose any point in this plane for the point o, the velocity-potential for all points in the plane will be zero; and for all other points will be proportional to the distance from this plane, being positive on the side towards which the body is moving, and negative on the other side.

EQUIPOTENTIAL SURFACES.

In general, *the equipotential surfaces are perpendicular to the lines of flow.* We have already seen that if we suppose the velocity of every point of the body to be marked down at that point, so as to constitute a permanent diagram of the state of motion of the body at a given instant, then the rate of change of the circulation from o to p, when p moves in the diagram with unit velocity, is the component along the tangent to the path of p of the instantaneous velocity at the point p. Hence if we now use P to denote the velocity-potential at p, viz. the circulation from o to p, we shall have $\partial_s P = v \cos \theta$, where v is the magnitude of the instantaneous velocity at p, and θ the angle it makes with the direction of s. Now those directions which lie in the equipotential surface through p are such that there is no change of potential when p moves along them, or $\partial_s P = 0$. Hence either $v = 0$ or $\cos \theta = 0$; that is, if there is any velocity, it is perpendicular to the equipotential surface.

If the motion of p is along a line of flow, $\cos \theta = 1$, and $\partial_s P = v$; that is to say, *the velocity at any point is the rate of change of potential per unit of length along a line of flow.* Hence if we take two equipotential surfaces very near to one another, the velocity at various points on one of these surfaces will be inversely proportional to the distance between them, with an approximation which is closer the nearer the surfaces are taken to one another. For the difference of velocity-potential between a point on one and a point on the other is constant; and the rate of change of P per unit of length is inversely proportional to the distance required to produce a given change in P.

Hence if we draw surfaces corresponding to the values 0. 1, 2,... of the velocity-potential, this system of surfaces will constitute a sort of diagram of the state of motion of the body. The velocity is everywhere at right angles to the equipotential surfaces, and where these are close together the velocity is large, where they are far apart it is small.

MOTION PARTLY IRROTATIONAL.

Suppose that in a mass of fluid there is a single vortex-ring of any form (i.e. a vortex-filament returning into itself), but that there is no rotation in any other part of the fluid. Consider a closed curve which is once linked with the ring, such as *abc*. The circulation round such a curve is equal to the circulation round a section of the vortex-filament, which we know to be the same for all sections; for the curve can be moved until it coincides with the section without crossing any vortex-line. Let the circulation round *abc* be called *C*.

We will now consider the circulation from a point *o* to a point *p*. Let the circulation along a path which goes from *o* to *p* entirely *outside* the vortex-ring be called (*op*). A path like *oxp*, which goes through the ring, can be altered, without crossing any vortex-line, into the form *orsr'p*, in which it is made up of a path *orr'p* outside the ring, and a path *rsr'* linked with the ring. Hence the circulation along *oxp* is made up of the circulations along these two paths, or it is (*op*) + *C*. A path such as *oyp*, which is twice linked with the ring, may be altered into a path going outside the ring together with *two* such closed paths as *rsr'*, and consequently the circulation along

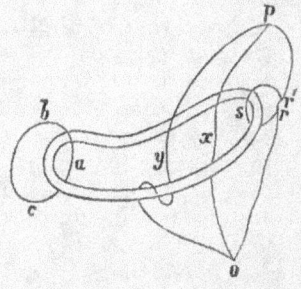

it is $(op) + 2C$. And generally, the circulation along a path which is m times linked with the ring, the same way round as rsr', will be $(op) + mC$. If it is linked with the ring by going the other way round it, the circulation will be $(op) - mC$.

Thus the circulation from o to p has not, as in the case of irrotational motion, a single definite value independent of the path pursued, but an infinite number of values included in the formula $(op) + mC$, where m is an integer number positive or negative. We may still speak of the velocity-potential at p, but it is now a *many-valued function* of the position p. We may compare it with the angle which has a given tangent; if θ be one value of the angle, there is an infinite number of other values, included in the formula $\theta + m\pi$, where m is an integer positive or negative.

If there are any number of vortex-rings, and the circulations round paths linked once with them are respectively $C_1 C_2 \dots$, then the circulation along a path from o to p linked m_1 times with the first, m_2 times with the second, etc., is $(op) + m_1 C_1 + m_2 C_2 + \dots$

In such cases we may still find equipotential surfaces. The equipotential surfaces through a point p contains all those points q which can be got at from p by a path along which the circulation is zero. Then the system of values of the velocity-potential at p is the same as the system of values of the velocity-potential at q.

Every equipotential surface meets every section of each vortex-filament, and breaks off there. Thus if there is only one vortex-filament, the equipotential surfaces partially consist of caps, covering the contour of the ring, as indicated in this figure. They must of course break off at the surface of the filament, because there is no velocity-potential inside the region where the motion is rotational.

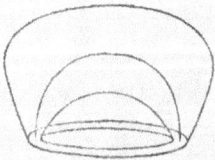

We say they *par-*

tially consist of such caps, because some of them may be in separate portions. When there are two or more vortex-rings, they may be joined by equipotential surfaces, as in this figure ; or an equipotential surface may consist of caps covering different rings, together with detached closed portions. 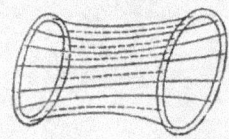 The proof of this proposition is very simple. Consider any section of a vortex-filament, and a point p on the boundary of the section. Choosing a single value P of the velocity-potential at p, let this vary continuously as p moves round the boundary of the section. Then it will gradually increase from P to $P + C$, where C is the circulation round the filament. Consequently every possible value of the velocity-potential will be represented on the boundary, and therefore every equipotential surface will meet it.

We may restore a one-valued velocity-potential by drawing caps to cover all the vortex-rings, and then defining the potential at p to be the circulation from o to p along a path which does not cross any of these caps. The caps are then called *diaphragms*. On the two sides of a diaphragm covering a vortex-ring the circulation round whose section is C, the velocity-potential will differ by C. Thus in crossing the diaphragm we should find a sudden jump in the velocity-potential. When a vortex-ring has two ends in the surface of the fluid, we must join these ends by any line in the surface, and then draw a cap covering the contour thus formed.

EXPANSION.

In general, the volume occupied by a finite portion of the moving body will increase or diminish in consequence of its motion. We proceed now to find, in the case of a homogeneous strain-flux, the rate of increase of a unit of volume of the body.

For the unit of volume we take the cube, three of whose sides are i, j, k. Let ol, om, on be unit lengths measured on the axes oX, oY, oZ. Then we shall have

velocity of $l = \phi i = ai + hj + g'k$,

    ,, $m = \phi j = h'i + bj + fk$,

    ,, $n = \phi k = gi + f'j + ck$.

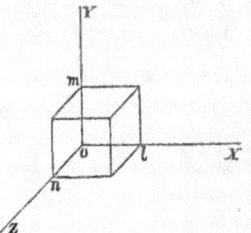

And in consequence of these velocities the cube $olmn$ will change itself into a parallel-epiped whose sides will be the new positions of ol, om, on. Now considering first the motion of l, we see that the volume of the cube will not be altered by any component of velocity parallel to the plane omn; because parallel-epipeds standing on the same base and between parallel planes are of equal volume. Hence the only part of the motion of l which can produce change of volume is the component of its velocity parallel to ol, which is a. And it is clear that the rate of change of volume due to this motion is precisely the velocity of l along ol, since the area omn is unity. Similarly the rate of change of volume due to the motion of m is b, and that due to the motion of n is c. As these three changes take place simultaneously, the whole rate of change of the volume is $a + b + c$. Volume is poured into the cube, as it were, through three faces of it, and so the whole change of volume is the sum of the changes due to the three faces. The quantity $a + b + c$ is called the *expansion*.

If we consider any other volume, containing V units, the rate of increase of that volume will be $V(a + b + c)$.

When the strain-flux is not homogeneous, we must divide the volume of the fluid into very small parts, one of which shall be called δV. Throughout this part the strain-flux is approximately homogeneous, and the expansion is $\partial_x u + \partial_y v + \partial_z w$, if the velocity $\sigma = ui + vj + wk$. Hence the rate of increase of δV is approximately $(\partial_x u + \partial_y v + \partial_z w)\,\delta V$, and consequently the rate of increase of the whole volume considered is $\Sigma\,(\partial_x u + \partial_y v + \partial_z w)\,\delta V$, with an approximation which is closer the smaller the parts δV are taken. Hence this rate of increase is exactly equal to the integral $\int (\partial_x u + \partial_y v + \partial_z w)\, dV$, since this also is represented by the same sum with an approximation which is closer the smaller the parts δV are taken.

We shall write E for the quantity $\partial_x u + \partial_y v + \partial_z w$, so that the rate of increase of a volume V is $\int E dV$. Of course, if the expansion E is a negative quantity at any point, the volume of the moving body is diminishing at that point.

From the value just found for the expansion it follows that if two velocity-systems are compounded together, the expansion at any point in the resultant system is the sum of the expansions at that point in the component systems.

The rate of increase of a finite volume of the fluid may also be calculated in another way. Consider a portion $\delta \alpha$ of the surface of that finite volume, so small that it may be regarded as approximately flat. Then if the fluid is flowing out of the volume at that part, there will be a rate of increase of volume equal to the magnitude of $\delta \alpha$ multiplied by the component of the velocity σ perpendicular to it; that is, equal to $-S\sigma \delta \alpha$. And this will be negative if the fluid is flowing into the volume. The whole rate of change of the volume will be the sum of the rates of change due to all the small parts of the surface, and is therefore equal to the integral $-\int S\sigma d\alpha$. Hence we have

$$-\int S\sigma d\alpha = \int E dV,$$

or, *the surface-integral of the velocity is equal to the volume-integral of the expansion.*

The scaler quantity E is derived from the vector σ, $= ui + vj + wk$, by the equation $E = \partial_x u + \partial_y v + \partial_z w$. If any vector be distributed over space so as to vary continuously from point to point, like σ, we might in the same way deduce a scalar quantity E from it. Prof. Clerk Maxwell calls the quantity $-E$ the *convergence* of σ. We might perhaps therefore call E itself the *divergence* of σ. In this language we have proved that the surface-integral of any vector is equal to the volume-integral of its divergence.

We have seen that the surface-integral of the *spin* over any closed surface is zero; the theorem just proved shews therefore that the volume-integral of its divergence over any region is zero. Since the region may be taken as small as we like, it follows that *the spin has no divergence anywhere.* We may in fact easily verify that if

$$2p = \partial_y w - \partial_z v,$$
$$2q = \partial_z u - \partial_x w,$$
$$2r = \partial_x v - \partial_y u,$$

then
$$\partial_x p + \partial_y q + \partial_z r = 0.$$

CASE OF NO EXPANSION.

The motion of the fluid is in the direction in which the velocity-potential increases. Hence if the velocity-potential is a *minimum* at any point, i.e. if it increases in all directions as we move away from the point, the motion is away from the point in all directions, and therefore there is positive expansion. Similarly, if the velocity-potential is a *maximum* at any point, so that it increases in all directions as we go towards the point, then the motion is towards the point in all directions, and there is compression, or negative expansion.

If there is no expansion, positive or negative, within the region bounded by a closed surface, the greatest and least values of the velocity-potential in that region must be on the surface. For since there is no expansion, there cannot be a maximum or minimum value inside the region.

If therefore the velocity-potential is constant all over the surface, it must be constant throughout the enclosed region, since the greatest and least values of it are now equal. In particular, if it is zero all over the surface, it must be zero throughout the enclosed region.

Suppose that the value of the velocity-potential is given at every point of a closed surface, and that there is no expansion and no spin anywhere inside the enclosed region. Then we can prove that only one velocity-system (or only one distribution of velocity-potential) is possible within the region. It is not proved as yet that *any* motion is possible with the given distribution of potential over the surface, and with no expansion and no spin inside; but we can shew that there cannot be *two* such motions.

For suppose that P is a velocity-potential having the given value at the surface of the region, and such that it gives rise to no expansion anywhere inside. Let also Q be a velocity-potential, satisfying the same conditions. Then $P - Q$ is a velocity-potential having the value zero all over the surface, and giving rise to no expansion inside. Therefore $P - Q$ is zero throughout the enclosed region, or Q is the same velocity-potential as P.

If the velocity σ at any point has the components u, v, w, so that $\sigma = ui + vj + wk$, then we know that $u = \partial_x P$, $v = \partial_y P$, $w = \partial_z P$. Moreover the expansion E is equal to $\partial_x u + \partial_y v + \partial_z w$, and therefore to

$$\partial_x . \partial_x P + \partial_y . \partial_y P + \partial_z . \partial_z P,$$

or, as it is conveniently written, $(\partial_x^2 + \partial_y^2 + \partial_z^2) P$. We have therefore proved that there can be only one solution of this problem: given the value of a function P all over a closed surface, to find its value at all points of the included region so that it may satisfy the condition

$$(\partial_x^2 + \partial_y^2 + \partial_z^2) P = 0.$$

We have not proved that there *is* any solution of the problem. That it should be applicable to the motion we have considered it is clearly necessary that P should vary continuously from point to point of the closed surface.

We may now extend this theorem. Instead of supposing that there is *no* spin and *no* expansion within the region, let us suppose the spin (confined to separate vortex-rings) and the expansion to be *known* at every point throughout the region; and let us consider the problem, having given the value of the velocity-potential at every

14—2

point of the surface, and the expansion and spin at every point of the enclosed region, to find the motion inside the region. We can prove, as before, that there is only *one* solution of it.

For let σ be the velocity at any point satisfying the conditions of the problem; and let also τ be a velocity satisfying the same conditions. Then $\sigma - \tau$ will be a velocity giving no expansion and no spin within the given region, and a constant value to the velocity-potential all over the boundary. Hence the motion, whose velocity-system consists of the velocity $\sigma - \tau$ at every point, is one in which the velocity-potential is constant throughout the region; that is, it is no motion at all. Hence $\sigma = \tau$ everywhere, or the velocity-system σ is identical with the velocity-system τ.

Suppose now that the closed surface expands indefinitely in all directions, and that the velocity-potential is zero for all points on it. Then we arrive at this conclusion : when the expansion and the spin are given at every point of space, and when the velocity-potential (and, therefore, also the velocity) approaches to the limiting value zero as we go away to an infinite distance in all directions, there is only one velocity-system possible. We shall now proceed to *find* this velocity-system, from the given expansion and spin, by describing certain ideal motions, out of which all continuous velocity-systems may be built up.

SQUIRTS.

Suppose that the lines of flow are straight lines diverging from a fixed point, so that the fluid is everywhere streaming away from this point; that there is no spin anywhere, and no expansion except at the fixed point. We propose to investigate this state of motion.

Because there is no spin anywhere, there is a velocity-potential P, and the equipotential surfaces cut the lines of flow at right angles. Since the lines of flow are straight lines passing through a fixed point s, the equipotential surfaces must be spheres having that point for centre. If we take two of these very near to one another, the normal distance between them is everywhere the same; but we have shewn that the velocity, at different points of an equipotential surface, is inversely proportional to the normal distance of a contiguous surface. Hence it follows that *the velocities are equal at equal distances from s.*

The tubes of flow are cones having s for vertex. Let sac be such a cone, and let it be cut at ab and cd by spheres having their centres at s.

The figure sab is similar to the figure scd; hence the areas ab and cd are to one another as sa^2 to sc^2.

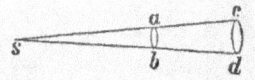

The area ab divided by sa^2 is called the *solid angle* of the cone at s. It is a *spherical measure* of the solid angle, just as the arc of a concentric circle, divided by the radius of that circle, is the *circular measure* of a plane angle.

Now the rate of increase of the volume $abdc$ is equal to the surface-integral of the velocity over its boundary. This boundary consists partly of the side of the cone and partly of the spherical ends ab and cd. The side of the cone can contribute nothing to the surface-integral, because at every point of it the direction of the velocity is in the surface, and consequently there is no component normal to the surface. The spherical surface ab, being everywhere perpendicular to the velocity, which is constant all over it, supplies a portion of the surface-integral, which is simply the product of the area ab by the velocity at any point of it, say a. Similarly the surface-integral due to cd is the product of the area cd by the reversed velocity at c. But since there is no expansion, these two must be equal and opposite in sign; therefore velocity at a × area ab = velocity at c × area cd; or, which is the same thing,

$$\text{velocity at } a \times sa^2 = \text{velocity at } c \times sc^2.$$

Or, we may say that the rate at which the fluid flows across ab, is the area ab multiplied by the velocity at a, and that if there is no expansion, the fluid must flow across cd at the same rate.

We learn thus, that in the motion considered, the velocity is inversely as the square of the distance from s. Let v be the magnitude of the velocity at distance r, then vr^2 is a constant, which we shall call μ. The circulation along a straight line sab from a to b is then

$$\int_{sa}^{sb} \frac{\mu}{r^2} dr = \frac{\mu}{sa} - \frac{\mu}{sb},$$

and hence if we make the velocity-potential zero at an infinite distance its value at distance r will be $-\dfrac{\mu}{r}$.

The rate of increase of any sphere of radius r, having its centre at s, is equal to the velocity at any point of its surface multiplied by the whole surface of the sphere. Now the surface of a sphere is $4\pi r^2$, and therefore the rate of increase is $4\pi r^2 v$, which is $4\pi\mu$. We should have expected this to be a constant, because there is no expansion in the space between two such spherical surfaces.

At the point s itself, the velocity-potential, the velocity, and the expansion, are all infinite, and we have no means of conceiving such a state of motion. To avoid this, we may imagine a very small sphere to be drawn round the point s, and the motion inside of this sphere to be replaced by a homogeneous strain-flux with the point s at rest, and the same velocity as in the original motion at all points on the surface of this sphere. The velocity will then vary continuously, and the motion will be conceivable at every point. Let E be the expansion of the homogeneous strain-flux, V the volume of the small sphere, then $EV = 4\pi\mu$.

The point s is called a *source* of strength μ when the fluid streams out in all directions; when μ is negative, so that the fluid streams inwards, it is called a *sink*. The whole velocity-system here described may be called a *squirt*.

WHIRLS.

Suppose next that the lines of flow are circles having their centres on a fixed axis, and their planes perpendicular to it, and that there is no spin except at the axis, and no expansion anywhere. Then the equipotential surfaces must be planes passing through the axis, and the velocity, being inversely proportional to the distance between two contiguous equipotential surfaces, must, for points on the same plane, be inversely proportional to the perpendicular distance from the axis. The condition that there shall be no expansion requires the velocity to be constant all round a circular line of flow. If the velocity at distance r from the axis be $\lambda : r$, where λ is a constant, the circulation round any line of flow will be the length of it, $2\pi r$, multiplied by the velocity $\lambda : r$; that is, it will be $2\pi\lambda$. The motion *at* the axis is inconceivable, as the velocity and the spin are infinite; but we may avoid this difficulty by drawing a very small cylinder round the axis, and supposing this to rotate about the axis as a rigid body, so that the points on its surface have the same velocity as in the original motion. This cylinder may then be regarded as an infinitely long straight vortex-filament, the circulation round any section of which is $2\pi\lambda$.

If we suppose a region of space to be marked out by a surface like the surface of a ring, and the axis to pass

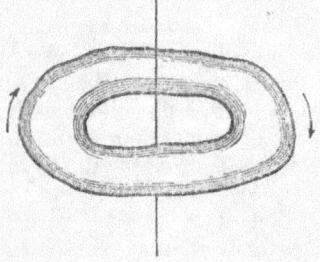

through the hole of this ring, but not into the region itself; then there will be no spin at any part of the region, and yet the fluid will flow continually round it. This explains how it is possible for fluid to flow continually round a re-entering channel, without ever having any motion of rotation. It is then possible to draw a closed curve within the region, which cannot be shrunk up into a point without passing out of the region. Whenever this is the case, it does not follow from there being no spin within the region, that the circulation round such a curve is zero; for it may, as in this case, embrace a vortex line lying outside of the region.

VORTICES.

We shall next investigate the motion in which there is no spin except at a certain closed curve, and in which the velocity-potential is proportional to the solid angle subtended by this closed curve at any point. By this we mean that from a point p lines are to be drawn to all points of the contour, forming a cone, and that this cone is to be cut by a sphere having its centre at p. The area which the cone marks off on the sphere, divided by the radius of the sphere, is the solid angle Ω subtended at p by the contour. Then the velocity-potential at p is $\nu\Omega$ where ν is constant.

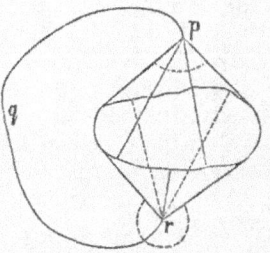

If we move the point p round the path pqr, the solid angle will diminish until it vanishes at some position near q. If we suppose a straight line passing through p to generate the cone, by moving round the contour in a definite direction, indicated by the arrows, the area on the spherical surface will be gone round in a definite way, by the intersection of the sphere with this moving line. We must then suppose the area on the left of the tracing point to be positive, and that on the right negative, p. 8. After the solid angle has acquired the value zero at q, it will change sign; and if we move our point on to r, the spherical area *inside* the cone must be reckoned negative. If we move on from r to p, passing through the contour, the area *inside* the cone at r will change continuously into the area *outside* the cone at p; and this is to be reckoned negative. Hence by going round a closed path which embraces the contour, we have continuously changed the solid angle Ω into $\Omega - 4\pi$. Hence the velocity-potential has by the same closed path been diminished by $4\pi\nu$, because it is changed from $\nu\Omega$ to $\nu(\Omega - 4\pi)$. It follows that the circulation round any path which embraces the contour is $4\pi\nu$, if the path go round in the direction $rqpr$.

If therefore we consider a piece of the contour so short as to be approximately straight, the motion in its immediate neighbourhood will be like that round the axis of a whirl for which $\lambda = 2\nu$. As in that case, we may draw a small tubular surface enclosing the contour, and substitute for the actual motion inside of it that of a small vortex-filament; so that any small length of this filament rotates like a cylinder about its axis. In this way we may make the velocity vary continuously, yet so that the motion is everywhere conceivable.

If we suppose the contour to be covered by a cap, and that the area of this cap is divided into a number of small areas, then the solid angle subtended by the contour at any point is the sum of the solid angles subtended at that point by all the small areas. Consequently the velocity-system just described, which may be called a *vortex*, is the resultant of a number of smaller vortices, whose vortex-lines are small closed curves which may be regarded as approximately plane. We shall now, therefore, examine more closely the case of such a small plane closed curve.

Take a point a within the area, and draw ax perpendicular to its plane. Let the angle $xap = \theta$, and the magnitude of the area $= A$. If we draw a sphere with centre p and radius pa, the area marked off on it by a cone with vertex p standing on A will be $A\cos\theta$ nearly. For if the area is small, the portion of the sphere cut by the cone may be regarded as approximately plane, and the generating lines of the cone are approximately parallel, 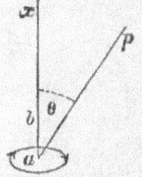 so that the spherical area is very nearly an orthogonal projection of A. Hence the solid angle subtended at p is nearly equal to $A \cos \theta : r^2$, if $r = ap$. Consequently the potential at p is $\nu A \cos \theta : r^2$ approximately.

Now we can produce the same potential in another way. Let us put at b a *source* of strength μ, and at a a *sink* of the same strength; then the potential at p due to this combination is

$$\frac{\mu}{bp} - \frac{\mu}{ap} = \mu \, \frac{ap - bp}{ap \cdot bp} = \mu \frac{bc \cos \theta}{r^2} \, , \text{ nearly.}$$

Hence if we make $\nu A = \mu \cdot bc$, and then let the area A and the length bc diminish continually, increasing ν and μ so as to keep νA, $= \mu \cdot bc$, $= k$, a finite quantity, both the vortex and the combination of a positive and negative squirt will continually approximate to the motion in which the velocity-potential is $k \cos \theta : r^2$. Now the source-and-sink combination gives no expansion except at a and b; consequently the limiting motion gives no expansion except at a. But we have seen that every vortex may be made up of component-vortices, whose vortex-rings are as small as we like. Hence these two conclusions :—

1. There is no expansion anywhere due to a vortex.

2. A vortex is equivalent to a system of squirts constructed in this way. Let two caps be drawn covering the vortex-ring, so as to be everywhere at a very small distance from each other. Let one of them be continuously covered with sources and the other with sinks, so that the source and sink on any normal are equal in strength, and so also that if $\mu \delta A$ be the total strength of the sources on the small piece of area δA, and t the thickness of the shell at that part, the product μt is constant all over the shell and equal to ν the constant of the vortex. Then, keeping all these conditions satisfied, this system of squirts will the more nearly approximate to the vortex the more nearly we make the two caps approach one another. For if we divide the cap into small areas, we have already seen that this is true of all the vortices whose vortex-rings are the boundaries of those areas.

Such a system of squirts is called a *double shell*. Inside the shell itself the velocity is not that due to the vortex, but is very large and in the contrary direction, namely, from the source to the sink. In crossing the shell the velocity potential is changed by $4\pi\nu$.

VELOCITY IN TERMS OF EXPANSION AND SPIN.

We are now able to resolve the velocity-system of an infinite mass of fluid, having no velocity-potential at infinity, into squirts and vortices.

Let E_a be the expansion at a point a. Suppose the entire volume divided into small portions, of which δV_a is the one including the point a. Place at the point a a source whose strength is $E_a \delta V_a : 4\pi$. Then the rate of increase of δV_a, due to this source, is $E_a \delta V_a$. And the velocity-potential at a point p, due to this source, is $-E_a \delta V_a : 4\pi r_{ap}$, where r_{ap} means the distance between a and p.

If a similar source be placed at some point inside each of the small pieces δV into which the volume is divided, the velocity-potential due to all of them will be $-\Sigma \dfrac{E_a \delta V_a}{4\pi r_{ap}}$. And if we indefinitely diminish the size and increase the number of the δV_a, this quantity will approximate to the integral $-\displaystyle\int \dfrac{E_a dV_a}{4\pi r_{ap}}$.

The meaning of the integral, however, requires examination. It supposes that every point where there is expansion is a source, so that in a region where the expansion is constant, the sources will be uniformly distributed. The strength of the source at each point must be zero, since the aggregate strength of all the sources in a portion of the volume is the rate of increase of that portion of the volume, which is finite. We have therefore to form the conception of a continuous distribution of source over a volume, so that the aggregate strength is a finite quantity, and yet there is a source at every included point. If, for example, sources are uniformly distributed in the interior of a sphere, the effect will be a homogeneous strain-flux, consisting of a uniform expansion of the sphere;

so that the velocity relative to the centre is proportional to the distance from it. When the distribution is variable, the *rate* of distribution at any point is what would be the aggregate strength of a unit of volume in which the distribution was uniformly what it actually is at that point. If S is the rate of source-distribution at any point, E the expansion at that point, then $E = 4\pi S$. For the rate of increase of a sphere of unit volume is $4\pi \times$ the aggregate strength of the sources within it.

We have given, then, a velocity-system in which the expansion at any point a is E_a, and the velocity-potential is zero at an infinite distance. We construct the system whose velocity-potential at a point p is $-\int \dfrac{E_a dV_a}{4\pi r_{ap}}$; and we shew that this system also has expansion E_a at every point a (for only the sources in the immediate neighbourhood of a point can produce expansion at the point), and its velocity-potential is obviously zero at an infinite distance.

If therefore in the given system there is no spin, the given and the constructed systems are identical; for if we subtract one from the other, we get a system in which there is no spin, no expansion (for the expansions in the two systems are everywhere the same), and no velocity-potential at infinity. And this, we have already proved, means no motion at all.

Next, let there be spin in the given system, and let Ω_p be the solid angle which a certain vortex-line subtends at p. Let a very small curve of area δA be drawn embracing the vortex-line, and let ω be the spin at that part; then $2\omega \delta A$ will be the circulation round this curve. Let $2\omega \delta A = \delta k$; then a motion whose velocity-potential at p is $\dfrac{\Omega_p \delta k}{4\pi}$ will have no expansion anywhere and no spin except at the vortex-ring. If we suppose the spin to be confined to isolated vortex-filaments, we may draw a surface across each filament, divide this surface into small areas δA, and draw a vortex-ring through some point in each one of these small areas. The sum $\Sigma \dfrac{\Omega_p \delta k}{4\pi}$ will

approximate to the integral $\int \frac{\Omega_p dk}{4\pi}$ if we indefinitely increase the number and diminish the size of the areas δA.
But this integral expresses a continuous distribution of
vortex-lines throughout the filament. If we suppose
straight vortex-lines to be uniformly distributed parallel
to the axis in a circular cylinder, the motion relative to
the axis will be rotation of the cylinder as a rigid body
about it. If then in the middle of the vortex-filament we
draw a very small vortex-filament of circular section, so
that a short piece of it may be regarded as a circular
section, the motion of this small filament will be compounded of two motions. First, an irrotational motion,
whose velocity-potential is $\int \frac{\Omega_p dk}{4\pi}$, calculated from all the
rest of the vortex-filament. Secondly, a rotation as of a
rigid cylinder about the axis with the spin ω.

Let us now suppose the integral $\int \Omega_p dk : 4\pi$ to be
extended over all vortex-filaments; whereby we may also
admit the possibility that these vortex-filaments continuously fill the whole of space. Then a motion constructed
so as to have this for its velocity-potential, and in places
where there is spin to be determined as just described, will
have everywhere no expansion and a spin equal to that of
the given motion.

If therefore we have given a motion in which E_a is the
expansion at any point a, and Ω_p the solid angle subtended
at p by a vortex-line, while the velocity-potential is zero
at infinity, and if we construct a motion having the ve-
locity-potential $-\int \frac{E_a dV_a}{4\pi r_{ap}} + \int \frac{\Omega_p dk_a}{4\pi}$, the two motions will
be identical. For if we subtract one from the other, we
shall get a motion in which there is no spin, no expansion,
and no potential at infinity; that is, no motion at all.

Thus we have shewn that if the expansion and the
spin are known at every point, the whole motion may be
determined. And the result is, that *every continuous
motion of an infinite body can be built up of squirts and
vortices.*

www.ingramcontent.com/pod-product-compliance
Lightning Source LLC
Chambersburg PA
CBHW071416170526
45165CB00001B/292